Laser Micro- and Nano-Scale Processing

Fundamentals and applications

IOP Series in Coherent Sources, Quantum Fundamentals, and Applications

About the Editor

F J Duarte is a laser physicist based in Western New York, USA. His career has expanded three continents while contributing in the academic, industrial and defense sectors. Duarte is editor/author of 15 laser optics books and sole author of three books: *Tunable Laser Optics, Quantum Optics for Engineers*, and *Fundamentals of Quantum Entanglement*. Duarte has made original contributions in the fields of coherent imaging, directed energy, high-power tunable lasers, laser metrology, liquid and solid-state organic gain media, narrow-linewidth tunable laser oscillators, organic semiconductor coherent emission, *N*-slit quantum interferometry, polarization rotation, quantum entanglement, and space-to-space secure interferometric communications. He is also the author of the generalized multiple-prism grating dispersion theory and pioneered the use of Dirac's quantum notation in *N*-slit interferometry and classical optics. His contributions have found applications in numerous fields, including astronomical instrumentation, dispersive optics, femtosecond laser microscopy, geodesics, gravitational lensing, heat transfer, laser isotope separation, laser medicine, laser pulse compression, laser spectroscopy, mathematical transforms, nonlinear optics, polarization optics, and tunable diode laser design. Duarte was elected Fellow of the Australian Institute of Physics in 1987 and Fellow of the Optical Society of America in 1993. He has received various recognitions, including the *Paul F Foreman Engineering Excellence Award* and the *David Richardson Medal* from the Optical Society.

Coherent Sources, Quantum Fundamentals, and Applications

Since its discovery, the laser, has found innumerable applications from astronomy to… zoology. Subsequently, we have also become familiar with additional sources of coherent radiation such as the free electron laser, optical parametric oscillators, and coherent interferometric emitters. The aim of this book Series in Coherent Sources, Quantum Fundamentals, and Applications is to explore and explain the physics and technology of widely applied sources of coherent radiation and to match them with utilitarian and cutting-edge scientific applications. Coherent sources of interest are those that offer advantages in particular emission characteristics areas such as broad tunability, high spectral coherence, high energy, or high power. An additional area of inclusion are those coherent sources capable of high performance in the miniaturized realm. Understanding of quantum fundamentals can lead to new and better coherent sources and unimagined scientific and technological applications. Application areas of interest include the industrial, commercial, and medical sectors. Also, particular attention is given to scientific applications with a bright future such as coherent spectroscopy, astronomy, biophotonics, space communications, space interferometry, quantum entanglement, and quantum interference.

Publishing benefits
Authors are encouraged to take advantage of the features made possible by electronic publication to enhance the reader experience through the use of colour, animation and video, and incorporating supplementary files in their work.

Do you have an idea of a book that you'd like to explore?
For further information and details of submitting book proposals, see iopscience.org/books or contact Ashley Gasque at ashley.gasque@iop.org.

Laser Micro- and Nano-Scale Processing

Fundamentals and applications

Edited by
Ahmed Issa
Engineering Department, Faculty of Engineering and Information Technology, Al Azhar University—Gaza, Palestine

Dermot Brabazon
I-Form, Advanced Manufacturing Research Centre, and Advanced Processing Technology Research Centre, School of Mechanical and Manufacturing Engineering, Dublin City University, Collins Avenue, Dublin 9, Ireland

IOP Publishing, Bristol, UK

ISBN 978-0-7503-1683-5 (ebook)
ISBN 978-0-7503-1681-1 (print)
ISBN 978-0-7503-2029-0 (myPrint)
ISBN 978-0-7503-1682-8 (mobi)

DOI 10.1088/978-0-7503-1683-5

Version: 20210801

IOP ebooks

British Library Cataloguing-in-Publication Data: A catalogue record for this book is available from the British Library.

Published by IOP Publishing, wholly owned by The Institute of Physics, London

IOP Publishing, Temple Circus, Temple Way, Bristol, BS1 6HG, UK

US Office: IOP Publishing, Inc., 190 North Independence Mall West, Suite 601, Philadelphia, PA 19106, USA

Cover image: Picture of a graphite sample surface microtextured via Nd:YAG pulsed laser ablation in liquid (PLAL) to produce various carbon nanoparticle types, sizes, and distributions. Image from work presented in chapter 8, Freeland *et al.*

To my beloved wife Aya, and precious children, Abdulraouf, Walid, and Alma

—Assistant Professor Dr Ahmed Issa

To my beloved wife Nicola, and precious daughter, Caoimhe

—Professor Dermot Brabazon

Contents

Preface

This book on laser micro- and nano-scale processing covers both the fundamental understanding and example usage laser system and applications. The book is presented in 11 chapters. After the introductory chapter, the range of laser systems, types and beam properties commonly utilised as well as recent developments in these technologies are presented in chapter 2. The physical principles of laser–material interaction are covered in chapter 3, and chapter 4 presents an overview of how laser processing parameters can be set for micro- and nano-scale processing.

Laser processing technologies have developed considerably over the last couple of decades. These developments are well highlighted by examples of recent laser applications as presented in chapters 5 through 11. A specific focus on surface processing at the micro- and nano-scales is presented in chapter 5 with application to generating surface areas which provide increased adhesive bonding for carbon fibre reinforced polymers. In a follow-on chapter, chapter 6, the effect of the laser generated chemistry and surface topology on the wettability and surface energy is covered.

Significant effort is represented in the published literature with a focus on trying to model laser–material interaction. The difficulty of modelling laser processing is evident from the complexity of the rapid physical phenomena that occur in multiple physical alteration regimes and scales. Chapter 6 presents an overview of the published modelling work that has been applied to predict the thermal field, the resulting ablated regions, and the resulting generated material microstructure.

The specific application of lasers to solar cell manufacturing in dye sensitised solar cells is presented in chapter 9. This includes laser surface treatment and nano-scale texturing for enhanced photo-absorption via both photothermal and photochemical processes. Chapter 10 presents an overview of recent developments in the laser micro-processing of polymers and silicon for application in microfluidics, and chapter 11 presents an overview of the latest research and application of both micro- and nano-scale laser surface structuring for a wide range of dental applications.

The audience for this work covers both academic and industrial researchers. It is important for the correct laser processing definitions, concepts, and state-of-the-art to be understood for its usage and further development. This book will therefore be an invaluable reference for engineers, scientists and computation researchers starting out or already working in the field of laser processing.

Each chapter is written by world experts in their area. As well as providing the latest background information, the state of the art in the niche applications areas is also presented. We take this opportunity to thank the 26 authors from across the world who have contributed the chapters. It has been enjoyable to work with you and encouraging to see your expertise, interest and desire to help others from your contribution. We also thank the IOP Publishing team who supported us in a professional manner in the compiling of this work. In particular, we thank Robert Trevelyan and Ashley Gasque on the IOP Editorial team and Francisco Duarte in his role as Series Editor, for their direction and support throughout the preparation of this book.

Dermot Brabazon and Ahmed Issa
June 2021

Editor biographies

Ahmed Issa

Dr Ahmed Issa, Assistant Professor is an Assistant Professor of Mechatronics Engineering, obtained his Bachelor of Mechatronics from International Islamic University, Malaysia in 2001, then obtained his PhD degree in Mechanical and Manufacturing Engineering (Lasers Micromachining) from Dublin City University, Ireland in 2007. His PhD research work focused on the analysis, control and simulation of micromachining processes of dielectric materials using Nd:YVO_4 and CO_2 laser systems. He joined Al Azhar University, Gaza, Palestine, in the year 2009. He is the former Assistant Vice President for Academic Affairs. He is currently the director of an Erasmus+ Capacity Building in Higher Education, EAC/A03/2016 project, entitled 'Increasing the Conformance of Academia towards Rehabilitation Engineering (i-CARE)'. He is the general chair of the 2020 International Conference on Assistive and Rehabilitation Technologies (iCareTech2020). He is a board member of the higher council for innovation and excellence, Palestine. In 2017, he received Zamalah fellowship funding to carry out research entitled 'Development of Engineering Laboratory Experiments for Shared and Remote Learning between Ireland and Palestine' at Dublin City University. He has research interests in Laser Processing of Materials and Thermal Mathematical Modelling, Mechatronics, Robotics, Measurements, Instrumentation, Control Systems, Automation and RP Technologies.

Dermot Brabazon

Full Professor Dermot Brabazon: BEng, PhD, is currently the Director for the Advanced Processing Technology Research Centre at Dublin City University (DCU), Deputy Director of I-Form, Advanced Manufacturing Research Centre, and is an academic staff member in the DCU School of Mechanical and Manufacturing Engineering. He has previous industry experience with Materials Ireland and many company support projects solving industrial related materials problems. An experienced project leader, Professor Brabazon has partnered in and led many national and international projects. These include Enterprise Ireland Innovation Partnership projects with Castolin Eutectic and Mincon Ltd; and the ACTTiVAte H2020 Pan European project for technology transfers and new value chain development. He is also very involved in the international scientific community as evidenced by his membership: on the Board of directors of the EU materials forming society (ESAFORM); on the conference Steering Committee for Advances in Materials and Processing Technologies (AMPT); in COST Action MP1401 Fibre Lasers; and his active collaborations and participation in Erasmus and Marie Curie programmes. Professor Brabazon has over 250 peer-reviewed publications (h-index: 33). Professor Brabazon was conferred with the President's

Award for Research in 2009 and the AMPT Gold Medal for life time service achievement in Research and Teaching in 2018. Professor Brabazon's research expertise is in materials and processing technologies, with a focus on laser processing, additive manufacturing, matrix composite production, and material phase structure analysis.

List of contributors

Yalda Afkham
I-Form, Advanced Manufacturing Research Centre, and Advanced Processing Technology Research Centre, School of Mechanical and Manufacturing Engineering, Dublin City University, Collins Avenue, Dublin 9, Ireland

Esther Titilayo Akinlabi
Pan African University for Life and Earth Sciences Institute (PAULESI), Ibadan, Nigeria

Ahmed Al-Hamaoy
College of Engineering, Al-Nahrain University, Baghdad, Iraq

Ahmad AlShaer
School of Engineering, University of Central Lancashire (UCLan), Preston, UK

Dermot Brabazon
I-Form, Advanced Manufacturing Research Centre, and Advanced Processing Technology Research Centre, School of Mechanical and Manufacturing Engineering, Dublin City University, Collins Avenue, Dublin 9, Ireland

Maila Castellano
Department of Chemistry and Industrial Chemistry, University of Genova, Genoa, Italy

Greg Foley
I-Form, Advanced Manufacturing Research Centre, and Advanced Processing Technology Research Centre, School of Biotechnology, Dublin City University, Collins Avenue, Dublin 9, Ireland

Brian Freeland
I-Form, Advanced Manufacturing Research Centre, and Advanced Processing Technology Research Centre, School of Mechanical and Manufacturing Engineering, Dublin City University, Collins Avenue, Dublin 9, Ireland

Carla Gambaro
Department of Mechanical Engineering, Polytechnic School, University of Genova, Genoa, Italy

Jennifer Gaughran
I-Form, Advanced Manufacturing Research Centre, and Advanced Processing Technology Research Centre, School of Physics, Dublin City University, Collins Avenue, Dublin 9, Ireland

Furat I Hussein
Mechatronics Engineering Department, Al-Khwarizmi College of Engineering, University of Baghdad, Iraq

Ahmed Issa
Engineering Department, Faculty of Engineering and Information Technology, Al Azhar University—Gaza, Palestine

Israt Kabir
Department of Engineering, School of Mathematics, Computer Science and Engineering, City, University of London, UK

Enrico Lertora
Department of Mechanical Engineering, Polytechnic School, University of Genova, Genoa, Italy

Chiara Mandolfino
Department of Mechanical Engineering, Polytechnic School, University of Genova, Genoa, Italy

Rasheedat Modupe Mahamood
University of Johannesburg, South Africa
Department of Materials and Metallurgical Engineering, University of Ilorin, Nigeria

Eanna McCarthy
I-Form, Advanced Manufacturing Research Centre, and Advanced Processing Technology Research Centre, School of Mechanical and Manufacturing Engineering, Dublin City University, Collins Avenue, Dublin 9, Ireland

Sumsun Naher
Department of Engineering, School of Mathematics, Computer Science and Engineering, City, University of London, UK

Verica Pavlic
Department of Periodontology and Oral Medicine, Medical Faculty, University of Banja Luka, Republic of Srpska

Marco Pizzorni
Department of Mechanical Engineering, Polytechnic School, University of Genova, Genoa, Italy

Shashi Prakash
Mechanical Engineering Department, Netaji Subhash University of Technology, New Delhi, India

Fatema Rajab
Laser and Optoelectronics Engineering Department, College of Engineering, Al-Nahrain University, Baghdad, Iraq

Bassam Rasheed
College of Engineering, Al-Nahrain University, Baghdad, Iraq

Tayf Sahib
University of Information Technology and Communications, Baghdad, Iraq

Sithara Sreenilayam
I-Form, Advanced Manufacturing Research Centre, and Advanced Processing Technology Research Centre, School of Mechanical and Manufacturing Engineering, Dublin City University, Collins Avenue, Dublin 9, Ireland

Silvia Vicini
Department of Chemistry and Industrial Chemistry, University of Genova, Genoa

Chapter 1

Introduction

Ahmed Issa and Dermot Brabazon

LASER stands for light amplification by stimulated emission of radiation. Since their invention, lasers have gained a wide range of applications in the fields of engineering, biomedical device production, surgery, dentistry, telecommunications, energy, automotive and aerospace, and other industrial fields. The first laser, a ruby laser, was reported in 1960 by T H Maiman [1]. Four years after that, some publications appeared reporting laser-induced damage in materials. Some of the first publications were those of Guiliano and Chiao *et al.* [2, 3] However, Chiao *et al* reported that this phenomenon was first reported by M Hercher [4]. These initial workers reported the generation of thin, long streaks of ionizations and damage due to focusing intense laser beams inside optical materials. Whereas low intensity laser beams pass through transparent materials without causing any observable effects, at sufficiently high laser intensities, absorption, refractive index changes, removal of the material from the surface, production of internal voids, melting, vaporization and even violent shattering may be induced.

Micro-/nanomachining or micro-/nanofabrication can be most easily defined as machining processes which result in structures that can be measured on micrometer and nanometer scales, respectively. Micro- and nanomachining can be applied to metals and other materials as well as semiconducting materials. In recent years, the structural alterations produced in materials by ultra-short laser pulses have been used for micro- and nanomachining. The availability of laser pulses with femtosecond duration enabled materials to be subjected to higher laser intensities than ever before, opening the door to the study of laser–material interactions in new regimes. Despite this long history, much still remains to be learned about the interaction of high-intensity laser pulses with materials.

Due to the unique properties of lasers such as directivity, monochromaticity coherence, and high-power irradiances or intensities, various technologies, applications, and processes were made possible employing lasers at research and industrial levels. Moreover, Q-switched laser pulses can be produced with high power, energy

doi:10.1088/978-0-7503-1683-5ch1

or fluence. Q-switching is achieved by storing energy during the population inversion (pumping) of the lasing material's atoms until it reaches a certain level and then releasing it very quickly. Q-switched or pulsed laser beams are more efficient for ablation due to the high energies that can be delivered during short pulse durations. This leads to more precise and confined energy deposition in the focal region. Furthermore, semiconducting materials can only be efficiently ablated using pulsed laser beams because they produce high peak intensities within short time durations. Many micro- and nanomachining applications in scientific research and industry have been developed around the use of lasers in the last few decades. Laser systems used for micro- and nanomachining include CO_2, Nd:YVO_4, Nd:YAG, excimer, Ti: sapphire, and fiber lasers.

Micro and nanomachining is achieved by tightly focusing a laser beam on a surface or underneath the surface of a target material causing localized heating, melting, and the subsequent ablation of the material in the focal region. Due to the tight focusability of lasers, the scale of the induced damage is generally in the micro and nanometer range.

The name and objective of micro- and nanomachining implies the necessity of using a machining tool that is smaller or at least comparable to the desired structure's scale. To accomplish that, two laser beam attributes are the most significant; the wavelength and pulse duration, as they both contribute to the machining scale and precision. Focusability is a direct function of the laser emission wavelength. In this field, the lasers used have wavelengths that range between few hundreds of nanometers such as the excimer laser (193 nm), or the third (355 nm) and fourth (266 nm) harmonics of Nd:YVO_4 or Nd:YAG lasers (1604 nm). Moreover, lasers with wavelengths of a few micrometers such as CO_2 lasers (10.6 μm) are also used for micromachining. Proper focusing of laser beams can lead to an effective focal spot size that is comparable to the emission wavelength. If tight focusing is employed via high numerical aperture objective lenses, then the focal spot can be downsized to fractions of the emission wavelength. The latter fact means that spot sizes of submicrometer or micrometer sizes can be achieved, which allows for micro- and nanomachining.

Another important factor is the laser pulse duration. The development of ultra-short pulse duration lasers has also contributed to enhanced machining precision. In this case, the absorption or diffusion timescales, over which the thermal energy is deposited or transferred to the target materials, are very small, resulting in confined and localized machining/ablation with minimal or no heat affected zones (HAZs). The short and ultra-short pulse duration advantage may be magnified by having a laser system that can deliver the pulses at a high repetition rate, which can ensure a uniform heat deposition and a reduced amount of thermal stresses or microcracking.

Other laser beam properties that make them superior in micro- and nano-machining include the absence of physical contact, which means that laser micro- and nanomachining does not involve physical contact between the tool and the processed material. This leads to more precise machining without mechanical forces between the sample and tool that can cause residual stresses, related post-machining defects, and eliminates tool wear as in traditional machining. Moreover, the intrinsic dimensional accuracy of laser processing is enhanced using computer control to guide laser beams leading to a high accuracy of heat deposition on to the target material. The latter has also allowed for machining miniature 2D and 3D intricate structures.

Laser micro- and nano-processing of metals includes processes such as ablation, hardening, shock peening, surface texturing, drilling, welding, cutting, and forming. The application could be the cutting of thin Nitinol metal to form stents, the heat treatment of tool steel drills for increased wear resistance and robustness, or the processing of dental amalgams for selective machining. There are a number of physical regimes which can occur in the laser–surface interaction region during laser processing. The temperature of the surface may be raised, a melt pool may be formed, and a plasma may be formed. The extent of superheating or plasma ionization is dependent on the amount of energy absorbed into the metal surface, which is dependent to a large extent on the material type, laser wavelength, laser pulse width, and fluence. A complex interaction exists between these inputs and a hyperspectral response surface is present for how these input parameters affect the formed part's properties. Therefore, this presents the basis of the important area of metal laser micro- and nano-processing research and development.

In the case of transparent materials, internal focusing allows for the fabrication (void forming) of micro- and nano-scale volume elements (voxels) inside the bulk of the transparent material. The internally fabricated structures may be continuous such as the fabrication of internal microchannels or waveguides. Since the micro-fabricated zone is either re-solidified material or void, it will have a refractive index different from the surrounding bulk of the material. This refractive index change is induced by the deposition of energy at that specific point. The fabricated structures are permanent due to the photo modification of the optical, mechanical, and chemical properties at the focal point. On the other hand, if the laser beam is focused on the surface, the micro-/nano-structurally altered zones are usually ablated zones.

There are also processes in which laser irradiation of liquidus materials causes solidification in the irradiated regions. This has opened the door for several applications in the fields of rapid prototyping and additive layer manufacturing. Such processes rely highly on computer control of a laser beam scanning system to produce the desired shapes on a layer-by-layer basis.

1.1 Book arrangement

The first technical focused review chapter of the book is chapter 2, with a presentation of the working principles and construction of laser systems, laser beam generation mechanisms, types of laser systems, properties of laser beams and a general preamble of laser areas of application. The applications of lasers in additive manufacturing and in the fabrication of micro- and nano-components are highlighted in the chapter.

Later, in chapter 3, laser–material interaction phenomena are presented for famous laser machining processes. Energy absorption mechanisms are explained with relevance to timescale effects, with description of mathematical and heat conduction models that govern the effects of laser irradiation of materials. The chapter gives a base-understanding for researchers on what to expect upon irradiating materials with laser beams. This understanding is delivered in terms of the laser–material interaction regimes and the most important parameters that govern these regimes leading to an understanding of what governs: the accuracy of micro- and nanomachining, minimized HAZ, how ablation or defects are induced, ablation thresholds, etc.

Chapter 4 introduces and discusses the influence of processing and laser beam parameters on the results of micro- and nano-laser machining applications. These parameters are categorized into laser beam parameters, material-related parameters, and processing parameters. The chapter, thereafter, gives a deeper discussion of definitions, nature, effects, mathematical models, and calculations related to each of these parameters. Furthermore, a simplified and critical presentation of workpiece scanning, processing and optical delivery system parameters is presented. Thus, the chapter forms an excellent and a comprehensive start for researchers who aim to plan and carry out experimental research in the field of micro- and nanomachining.

Chapter 5 aims to introduce and describe in depth the process of surface nano-structuring of metallic, polymeric, and composite materials by using different laser sources. Laser-induced surface texturing at the micro- and nano-scale is a new technology in the service of surface engineering, with the possibility of creating very fine structures with extremely high resolution. The superficial material changes focused on are wettability properties and modification of surface roughness. The chapter also gives an analysis of how these changes increase the adhesion properties of surfaces for applications in the field of structural bonding. Specific sections of the chapter discuss changes in the durability behavior of adhesive-bonded joints.

Chapter 6 aims to provide a general and concise overview of surface characterization techniques, which are useful to understand the effect of laser surface texturing, both on the micro- and nano-scale. The chapter begins with describing wettability theories and their link to the results of experimental investigations on laser surface texturing. The chapter then focuses on methods of investigation that allow qualitative and quantitative detection of superficial morphology changes. Finally, attention is given to investigations concerning the possible chemical modifications, due to oxidation states or any type of interactions of the surface layers, with the surrounding environment.

Chapter 7 focuses attention on a case study of laser surface glazing (LSG), which is a widely used process for improving surface hardness and wear resistance of steel. The chapter presents the development of a simple and reliable model of LSG for metallic materials using a finite element method. The model describes both 2D and 3D transient thermal energy distribution as a result of LSG with cylindrical geometry. The model predicts the temperature distributions, heating, cooling rates, and depth of the modified zone.

In chapter 8, pulsed laser ablation in a liquid process is presented. In this process a laser is fired at the interface between a solid and liquid, and a plasma is created from the solid, which in turn condenses back into nanoparticles retained within the liquid. This process can be used within a static liquid or a dynamic flow regime. In the static liquid scenario, a limit of production is hit after which the cloudiness of the colloid prevents the light energy from effectively reaching the surface. Within the flow regime, a continuous production of colloid can be achieved where both nanoparticle size and colloid concentration can be controlled. A wide range of applications is presented within this chapter for the highly defined resulting inks, including sensors, anti-fouling surfaces, therapeutics, and conductive circuits.

Chapter 9 gives a review on the potential application of laser surface processing in the fabrication of dye-sensitized solar cells (DSSCs). Conventional techniques used in the synthesis of DSSCs are presented, followed by a presentation of laser surface treatment as an alternative method that can be used to fabricate DSSCs. Focus is given to the advantages of laser processing such as localized and fast processing compared to conventional methods. The chapter reviews several research efforts and concludes that the use of a laser as a tool for modifying DSSCs is still under development, and a deeper understanding of the phenomena involved is much needed given its potential.

Chapter 10 presents a review and a case study of laser micromachining for fabricating microfluidic devices. The chapter focuses on the use of lasers as microfabrication tools widely used in the field of micro-electro-mechanical systems. The fundamental principles of various types of laser processes utilized for the micromachining of polymers and silicon are presented. The chapter then focuses in depth on presenting research findings and the challenges of using a CO_2 laser to fabricate microfluidic devices in polymers and silicon substrates.

Chapter 11 presents an overview of the use of lasers for nano- and micro-scale processing of materials for applications in modern dentistry. Surface bonding of dental implants, their osseointegration, and how the surface texture can be modified to achieve these are presented in this chapter. The direct implementation of lasers for tissue bonding and the laser additive manufacturing of implants are also presented as advanced laser structuring technologies being applied in modern dentistry.

1.2 A message from the editors

This book aims to make a valuable collection of chapters summarizing the collective authors' years of research achievements and knowledge gained in the field of laser nano- and micro-processing of materials. We hope that the book will fill a gap in the literature and save a lot of precious time spent on referring to several sources that address laser material processing. This text provides a sufficient and substantial starting point for young researchers at the beginning of their careers; and interesting recent knowledge gained for intermediate and advanced researchers in one text. We believe, therefore, that the book will make an excellent reference source for many research groups that have to train young researchers within a short time.

We wish you joyful, fruitful reading and successful research endeavors in the field of laser micro- and nanomachining.

Professor Ahmed Issa
Professor Dermot Brabazon

References

[1] Lippert T 2004 Laser application of polymers *Adv. Polym. Sci.* **168** 51–246

[2] Giuliano C R 1964 Laser-induced damage to transparent dielectric materials *Appl. Phys. Lett.* **5** 137–9

[3] Chiao R Y, Garmire E and Townes C H 1964 Self-trapping of optical beams *Phys. Rev. Lett.* **13** 479–82

[4] Hercher M 1964 *J. Opt. Soc. Am.* **54** 563

Chapter 2

Laser systems, types and beam properties

Rasheedat M Mahamood and Esther T Akinlabi

The term LASER stands for light amplification by simulated emission of radiation. Laser beams are generated from a light source that is then amplified via an iterative feedback loop in a way that is similar to the way a microphone system amplifies sound. The amplification of the light is achieved by a process called simulated emission which is also known as optical amplification. Laser light is characterized by a single wavelength, also known as monochromaticity, is coherent (of same phase) and has low beam divergence. These characteristics of a laser are responsible for its application in various human endeavours. The advent of the laser has helped to revolutionize how products are being manufactured. The much sought after miniaturization of equipment and consumer goods is made possible with the help of a laser that makes it possible to fabricate parts at the micro- and nanoscale levels. The laser has also allowed the repair of high-value components, which in the past were not possible to repair. This chapter presents a detailed working principle of a laser system, the properties of a laser, the different types of lasers and their areas of application. The applications of lasers in additive manufacturing and in the fabrication of micro- and nanocomponents are also highlighted.

2.1 Introduction

Laser sources used in industry are of coherent monochromatic radiation most often in an optical region of 0.2–20 μm wavelength. In the 1960s, the first laser, called the 'ruby laser' was made by Theodore Maiman but the actual idea of lasing was first explained by Albert Einstein, which was also based on the work of Max Planck from the late 18th century [1–5]. Einstein described the behaviour of atoms in different energy levels. He recognized that photon emission could be produced from an atom in two different ways. The first is spontaneous emission of photons occurring without external intervention. Second, an external photon can stimulate an atom or molecule to emit light if the energy of this incoming photon exactly matches the energy that an electron would release when dropping to a lower energy level. 'Laser'

doi:10.1088/978-0-7503-1683-5ch2

is an acronym (light amplification by simulated emission of radiation) that describes the principle of production of the coherent type of light. The principle of operation of a laser, the properties, the different types and the areas of application are presented in this chapter.

2.2 Working principles of lasers

The theory of quantum mechanics explains how electrons of atoms can take different states of energy levels such as E_1, E_2 and E_3, where E_1 is less than E_2, and E_2 is less than E_3. The principle of lasing involves many processes that include: simulated absorption, spontaneous emission and stimulated emission. These three processes take place in a two-level energy system. Let us consider placing an electron inside a cavity of a two-energy system as seen to occur inside an atom or molecule. Electrons are happy to stay at a lower energy level, but when a light is shone on the electron, it moves to a higher energy level. This excitation is caused as a result of the resonance of the incoming photon that is at the same frequency as the energy gap ΔE between the two energy levels as shown in figure 2.1. This process is called absorption. The excited electron cannot remain in the higher energy state for long and it typically exists at the higher energy levels for 10^{-8} s. After this time, the excited electron returns back to the lower energy level by releasing a photon. This process is referred to as spontaneous emission as demonstrated in figure 2.2.

Einstein thought that if the electron in an excited state comes in contact with an incoming photon, it will stimulate the excited electron to lose a photon and return

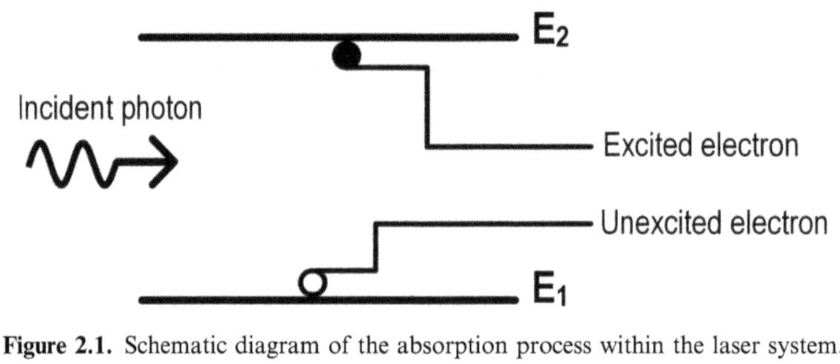

Figure 2.1. Schematic diagram of the absorption process within the laser system.

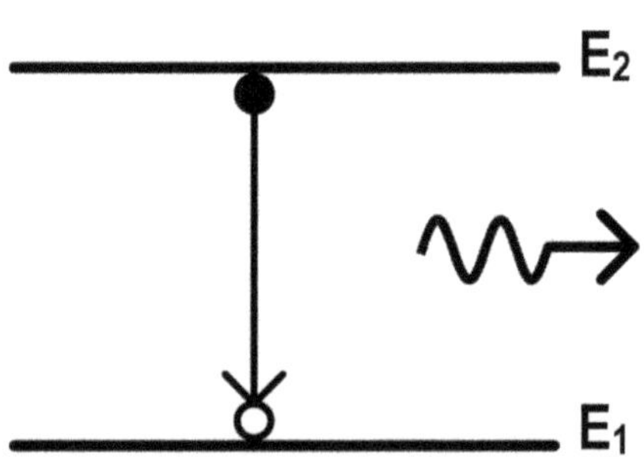

Figure 2.2. Spontaneous emission of a photon of energy as the electron returns for E_2 to E_1.

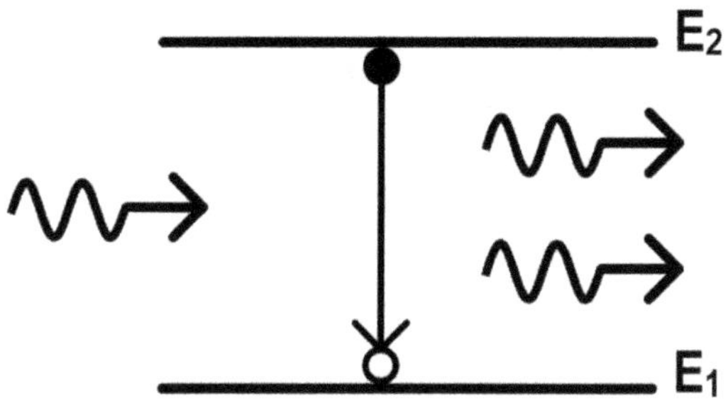

Figure 2.3. Schematic illustration of stimulated emission of photons after stimulation.

back to the lower energy state. The emitted photon and the incident photon are totally coherent and at the same frequency. This process is known as stimulated emission. What Einstein proposed is that instead of allowing spontaneous emission to occur, which produces photons of irregular frequencies and out of phase, a coherent photon can be produced through a stimulated emission process which is an important lasing process. This process is shown in figure 2.3. The rate of transition of electrons from the low energy state to high energy state (absorption rate) is given by the rate at which the number of electrons in the higher energy state increases, which is also proportional to the number of electrons in the lower energy level and energy density of the photons. Also, the rate at which the electrons in the higher energy state is depopulated is proportional to the spontaneous and stimulated emissions.

In a three state energy system, there are four processes involved in the production of a laser namely: absorption, spontaneous emission, pumping and population inversion and stimulated emission of electromagnetic radiation. The excited electron can only stay in the higher energy level for approximately 10^{-8} s; some materials can exist at this high energy state for longer and they are called active system or active media. Consider an electron having a three energy state in an active media as shown in figure 2.4. The electrons are very happy at the lowest energy level E_1. If a photon is introduced from an external source into the system and the photon is at the same frequency with the energy gap E_3–E_1, the electron will become excited by absorbing the photon and move to the highest energy level (the absorption process). The excited electron will return back to the lowest energy level after losing energy in the form of a photon after the lifetime elapse through the spontaneous emission process. While the electron in the high energy state can decay spontaneously, some of the electrons can also decay through a process that is called non-radiative decay. Some electrons will lose energy through collisions with the surrounding atoms, molecules or walls of the cavity. This loss of energy does not lead to emission of photons and the energy difference can change into kinetic energy or internal energy or thermal energy. Such electrons will undergo non-radiative decay to the intermediate energy level E_2. The intermediate energy level is also known as the metastable state. Electrons can stay longer in this metastable energy state and the existence of these electrons is achieved through what is known as the optical pumping process.

The back and forth movement of photons inside the gain cavity of the lasing medium is between two parallel mirrors [6]. The optical pumping is produced by a

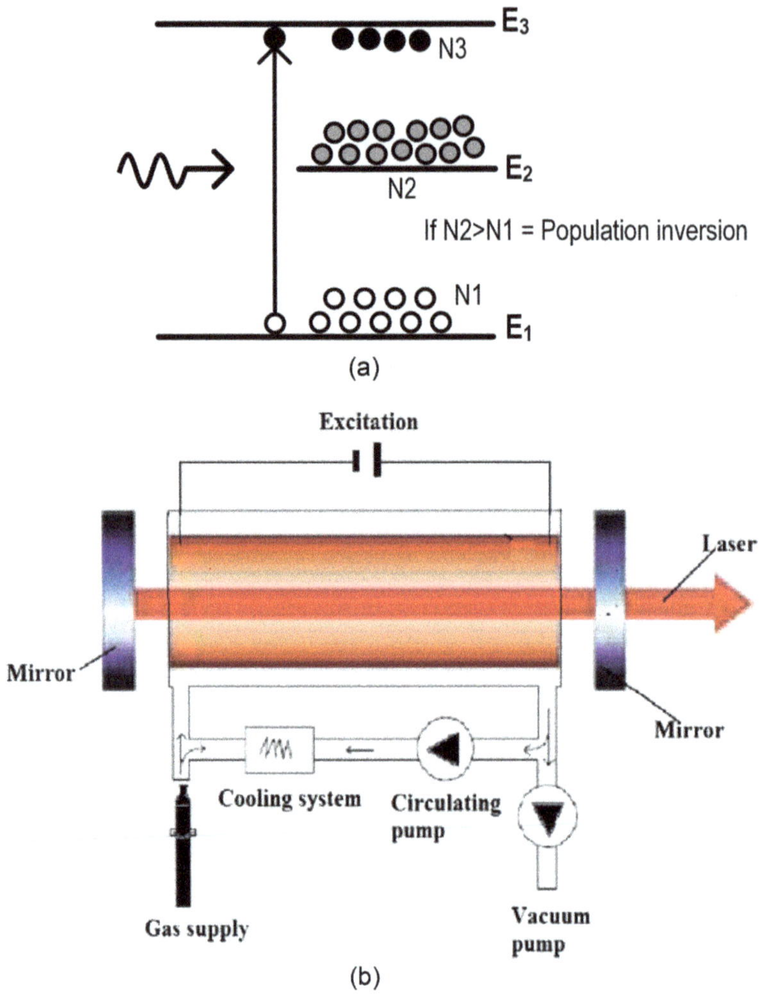

Figure 2.4. Schematics of (a) the population inversion and (b) optical pumping processes [6]. Copyright (2018), with permission of Springer.

supply of adequate energy into the active medium also known as the gain cavity, from an external source. This causes the absorption of photons by electrons at the lowest energy level and moves to the highest energy level, the spontaneous decay, the non-radiative decay and hence the population inversion. Let the number of electrons in E_1, E_2 and E_3 energy levels be N_1, N_2 and N_3, respectively, as shown in figure 2.4. If N_2 is greater than N_1, then population inversion has occurred. Usually, the population of electrons in the lower energy level E_1 is larger than those at higher energy levels. With the supply of energy from the external source (pumping), the processes of absorption, spontaneous emission and stimulated emission occur simultaneously in the lasing chamber. The absorption process dominates all the processes; hence the incident electromagnetic radiation cannot be amplified. The amplification of the incident electromagnetic radiation can only be achieved if and only if the population of electrons in the higher energy levels is greater than those at the lower energy levels, which is known as population inversion. This population inversion will allow the production incident light amplification due to the great number of stimulated photos produced during the stimulated emission process of the electrons in the metastable state. Population inversion is achieved by the pumping of atoms into the metastable level at a higher rate than the rate at which they leave this level. The pumped atoms need to stay in this metastable level without de-exciting or only a few de-exciting while the population keeps increasing. Population inversion is an important process that is required for lasing to occur in any laser system. When the incident photon causes the excited electron in the higher energy level to emit a photon that is at the same frequency with the incident photon and at temporal coherence, this causes the incidence photon or light to be amplified. This optical amplification is known as a laser that is coherent and also in phase. Simulated emission is crucial if lasing is to occur. The emitted photon is in phase with the incident photon, it has the same wavelength as the incident photon and it also travels in the same direction. The components of a laser system are explained in the following section.

2.2.1 Laser components

There are three basic components in a laser system, namely: the lasing medium, excitation or energy pumping mechanism and the optical cavity, shown in figure 2.4(b). The lasing medium is a substance that emits light in all directions when excited by energy. This substance could be a gas, liquid or semiconducting materials. The excitation or pump mechanism is the source of energy that is used to excite the lasing medium. The excitation mechanism that is used in lasing includes: electricity, flash tube, lamps or energy from another laser. The optical cavity consists of two mirrors, one at each end of the cavity or lasing medium. One of the two mirrors is 100% reflective while the other one is partially transparent, which allows the passage of the laser. The optical cavity is also referred to as the gain medium where the lights are reflected back and forth. As the light bounces between these two mirrors, the strength of light increases, which helps in the light amplification from the excitation mechanism. The output or the partially transparent mirror allows some of the light to leave the optical cavity which is then used for the production of the laser beam while only a portion of it will be reflected back into the cavity. There are different types of lasers which are discussed in the next section.

2.2.2 Types of laser

There are different types of laser and depending on the state or type of the lasing medium there are solid-state lasers, liquid-state lasers, gases and semiconductor lasers. A solid-state laser uses materials which are solid at room temperature. The lasing material is distributed in the matrix of solid materials. An example of this type of laser is the first ever made ruby laser, neodymium: yttrium–aluminium garnet 'YAG' laser. The neodymium–YAG laser emits infrared light at a wavelength of 1064 nm [2]. In a solid-state laser a crystalline solid or glass rod material is doped with ions such as neodymium (Nd) to produce the lasing medium. This type of laser is typically pumped using a flash tube or another laser with a shorter wavelength than the lasing medium wavelength and a solid-state laser can produce high powers in the infrared light spectrum [6]. Solid-state lasers generate a very high temperature in the lasing medium due to the excess pump power that heats up the lasing medium and which in turn reduces the quantum efficiency of the laser [6]. Apart from the ruby laser, other types of solid-state lasers include: the YAG-based lasers such as neodymium-based YAG laser—Nd:YAG laser; thulium-based YAG laser—Tm:YAG laser; ytterbium-based YAG laser—Yb:YAG laser; and the holinium YAG laser—Ho:YAG laser [2, 6].

The liquid-state laser is another type of laser in which the lasing medium is liquid at room temperature [7]. A dye laser is an example of a liquid-state laser. The liquid in a dye laser contains organic dye molecules that can emit light over a range of wavelengths. A laser with an adjustable wavelength can be produced using dye laser. This can be achieved by adjusting the laser cavity, which will in turn change the output wavelength of the laser. Complex organic dyes, such as rhodamine 6G, in liquid solution or suspension are used as lasing media in dye lasers that are tunable over a wide range of wavelengths. The spectrum ranges from the near-ultraviolet to near-infrared radiation depending on the type of dye that is used and the dye is usually doped into the liquid crystal to produce a continuous spectrum of lasing [7]. Arc lamps, flash lamps, or other types of lasers can be used to pump liquid lasers. The liquid-state laser can be operated in a pulse mode. This type of laser suffers instability due to high heat intensity that changes the refractive index of the active substance and then degenerates the rays in the lasing medium [6].

Another type of laser uses different gases as lasing media. An example of this type of laser is the helium–neon laser that can emit red, yellow, orange, green, or infrared light. Other types of gas laser include: helium–silver (HeAg) and neon–copper (NeCu) laser, argon laser, krypton laser, xenon ion laser, nitrogen laser, carbon dioxide laser, carbon monoxide laser and excimer laser. Excimer lasers are produced using reactive gases, such as chlorine and fluorine and mixed with inert gases such as argon, krypton or xenon. When excimer lasers are excited, a pseudo molecule (dimer) is produced and when lased, the dimer produces light in the ultraviolet range wavelength. Chemical lasers are another example of gas lasers. A chemical reaction between gases is used to produce a large amount of energy that can be released very quickly [2, 4]. Chemical lasers can produce a high level of power up to the megawatt power range. Examples of chemical lasers include the oxygen iodine laser and the deuterium fluoride (DF) laser.

Semiconductor lasers are sometimes called diode lasers and non-solid-state lasers. The lasing medium in a semiconductor laser is made up of a p–n junction between two regions that is doped with different materials that make up the semiconductor [8]. The lasing is achieved by electrically pumping the lasing medium, which is the interband transition under the conduction band, and the light oscillates in the junction plane. The optical gain in a semiconductor laser is achieved through the recombination of electrons and holes produced by the electrical pumping activities. Semiconductor lasers are very flexible and can be made to any required size and the wavelength depending on the type of semiconductor compound used. The properties of lasers are presented in the next section.

2.2.3 Laser beam characteristics

Laser light has three main characteristics that differentiate it from an ordinary light. These three characteristics are collimation or directionality, monochromaticity and coherency.

2.2.3.1 Collimation/non-divergence/directional

A laser beam is a collimated or a non-divergent beam of light, as depicted in figure 2.5. The process of producing a laser is what is responsible for this unique property of the light. During the stimulated emission process, the photons produced by the stimulated electron are in phase and at the same frequency with the incident photon. The many incident photons will be amplified and after stimulated emissions to produce electromagnetic waves travelling in a direction parallel to one another other and in a single direction with very little divergence. This characteristic of lasers is what makes it possible for laser beams to be focused on a specific area with very high intensity; as compared to ordinary light waves which spread out and lose their intensity as the distance from the source increases.

2.2.3.2 Monochromaticity

The process of lasing makes it possible to produce laser beams with a characteristic single wavelength that is referred to as monochromatic light. The property originates from a stimulated emission process in the resonant cavity and the laser medium which allows only about 1% of the stimulated light to pass through the partially reflecting mirror forcing the beam through a very narrow wavelength. Laser light has a very pure wavelength as a result of its manufacturing process. A laser beam is said to be monochromatic, which refers to the single (wavelength) colour of a laser beam making it possible to deliver laser energy to a specific area as needed; in contrast to ordinary white light, which is a combination of many different wavelengths (colours) as shown in figure 2.6.

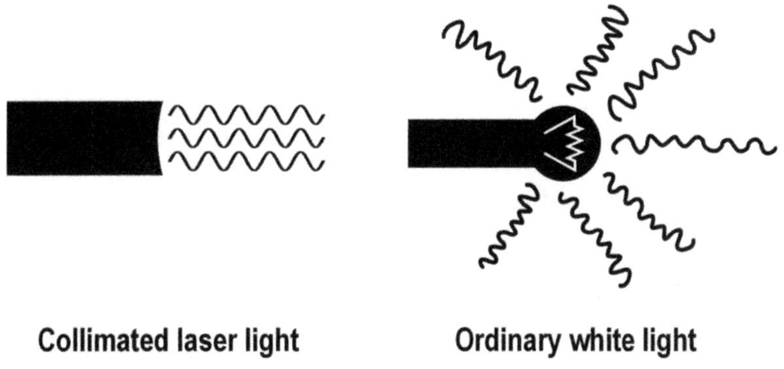

Figure 2.5. Collimated laser light and un-collimated ordinary white light.

Figure 2.6. (a) Monochromatic laser light and (b) non-monochromatic ordinary white light.

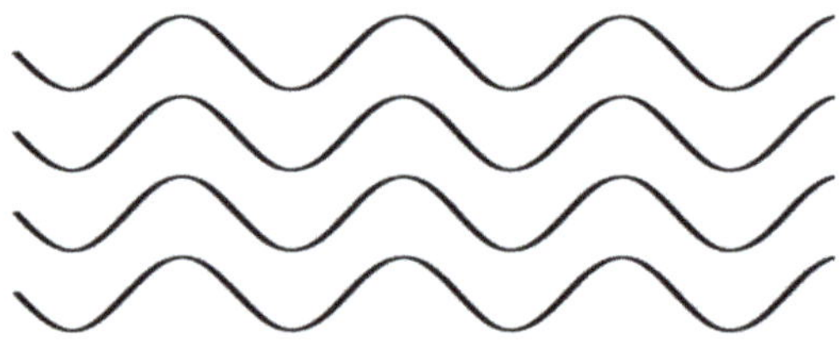

Figure 2.7. Laser light waves that are coherent.

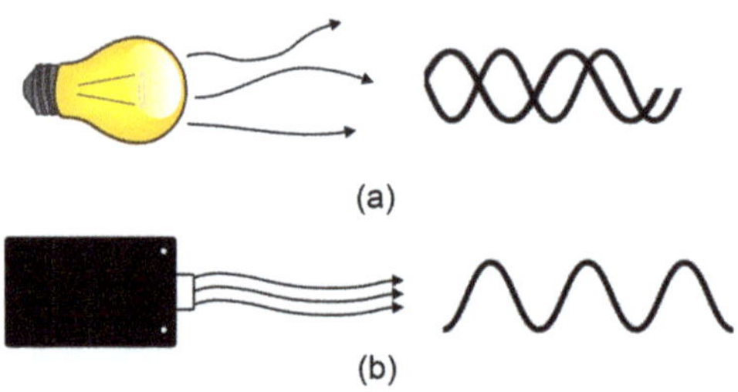

Figure 2.8. Schematic illustration of (a) incoherent and (b) coherent laser light.

2.2.3.3 Coherency

Coherency of laser light is achieved from the stimulated photon that is produced during the lasing process. The stimulated photon and the incident photon are at the same frequency and they are also in phase. This is why a laser light is coherent (see figure 2.7). All the electromagnetic light waves that are produced during lasing move together in phase, in both time and space. A laser has a very coherent beam that is strong and concentrated while an ordinary white light releases light in many directions and the waves are not in phase, as shown in figure 2.8.

2.3 Laser areas of application

The advent of lasers has really revolutionized our world. It has helped us to change the way components are made and joined. The properties of the laser make it able to deliver coherent, monochromatic and well-controlled directional light beams. Lasers have applications in many fields of human endeavour. These application areas range between low energy ranked lasers that are used as pointing pens in presentations, to very high energy rated lasers that are used in material manufacturing. Some of the areas of application of lasers are presented in this section.

2.3.1 Entertainment

Lasers are applied in the entertainment industry for projection and display of videos, laser lighting displays and laser light shows for cinemas. They are also used in home videos, flight simulators, and so on. Lasers find their application in optical data storage systems such as in compact discs, DVDs, Blu-ray discs and magneto-optical discs that rely on a laser source for their usage. The high spatial coherence property of lasers, which gives them the ability to be directed to a very tiny spot on a recording medium, makes it possible for them to be used for high density data storage devices. Optical discs are made of semiconductor lasers and are used for reading data from a variety of optical devices for playing music and video recording.

2.3.2 Consumer products

Laser printers, barcode scanners, laser pointers and holograms are some of the uses of lasers in consumer products. The properties of lasers that allows the beam to be focused on a very tiny spot and also to be put on and off at will have made lasers useful for these products. Barcode scanners have made the life of supermarket workers easier. By moving packages across a laser beam where the optical sensors detect the reflected light from barcodes on a package and relay it to a computer for adding the price of an item automatically is one of the interesting applications of lasers. Since its inception in 1974, the laser scanner for reading barcodes has become commonplace in supermarket and retail stores.

2.3.3 Military and energy applications

The ability of lasers to concentrate extremely high power has made them useful for military and energy applications. Nuclear weapons and missile defence are some of the uses of lasers in the military. High energy lasers provide a way to deliver destructive energy on a target and at a very fast rate for fast moving missiles. Lasers are also used by the military as weapons on land and at sea. Lasers are used as weapons to cause skin burns and retinal damage. The ability of lasers to provide the necessary extremely high temperatures to fuse atomic nuclei together in order to release energy has made them useful in energy industries. High power lasers can also be used for isotope separation.

2.3.4 Medical applications

The medical industries use lasers widely for a number of medical procedures and treatments. Lasers are now used for eye surgery and vision correction. Lasers of wavelengths close to one micrometre can be used to weld a detached retina in a human eye, or cut internal membranes in cataract surgical operations. Low power pulse lasers can be used to destroy abnormal growth that spreads across the retina in diabetic patients. Lasers are also used to remove tissue from the cornea to reshape the transparent outer layer of the eye. Dentistry also makes use of lasers for procedures. Dermatologists are widely using lasers for photodynamic therapy of cancer and other skin procedures in cosmetic treatments such as tattoo and hair removal. Lasers can be guided through optical fibres and used to activate photosensitive drugs that are absorbed by cancer cells to selectively destroy tumours. Lasers are used in surgery to cut tissues using laser radiation. Lasers are used for imaging such as in X-ray, ultrasound, etc. They have revolutionized the medical industries as they have changed the way human bodies are dissected for surgeries. With the use of lasers, finely controlled and relatively bloodless dissection and tissue-destroying procedures can be effectively carried out.

2.3.5 Measurement, alignment and imaging

Lasers find their application in measurements and alignments. The helium–neon laser was the first type of laser to find a wide range of commercial applications. The reason for this was that the laser could be adjusted to produce a visible red beam of laser light. Lasers are used for projecting straight lines for alignment, or as measuring tools, in many fields where measurements are required. Lasers have made the work of surveying easier. Surveyors and construction workers now use lasers to draw straight lines for their work. Surveyors bounce the beam off a mirror to measure direction and angle. The coherence property of laser light is what makes it important in interferometry and holograms that depend on light waves to make precise measurements and to record three-dimensional images. Lasers are also used in optical metrology and remote sensing applications.

2.3.6 Research tools' application

Lasers have proved very valuable for carrying out research precisely. Spectroscopy is an important area where the laser's properties have been put to good use in research. The study of the light absorbed and emitted in atoms and molecules helps to reveal the inner behaviour of atoms. Laser spectroscopy is used in a wide range of applications such as atmospheric physics and pollution monitoring, solid materials analysis and various types of fundamental research related to measurements. Microscopy is another important area where lasers have been used as research tools. Laser microscopes are used to obtain images at higher magnifications and at very high resolution.

2.3.7 Communication application

Lasers are applied in communication systems such as fibre-optic coupled semiconductor lasers and used to transmit long distance signals. The fibre-optic system is the backbone of any communication industry as a result of a number of benefits it offers that include: high data capacity, prevention of electromagnetic interference and improved security.

2.3.8 Industrial and manufacturing applications

Lasers have revolutionized the way products are made in the manufacturing industries and are a driving force for the fourth industrial revolution. The use of lasers in material processing ranges between altering the material properties for specialized application to the manufacturing of three-dimensional end-use parts [6, 9–11], to the fabrication of delicate parts, to the repair of high-value parts that were not previously repairable [12]. Lasers are widely used in manufacturing processes such as: laser cutting, laser drilling, laser welding, laser cladding, laser surface treatment, laser ablating, laser marking and engraving and laser micromachining. All these are non-contact manufacturing processes that have helped to offset some of the problems of the traditional manufacturing processes [13]. Lasers are the backbone of the much needed miniaturization drive that requires products to be made compact and lighter. Laser material processing [14–16] methods allow the fabrication of components at micro- and nanoscale levels with high quality and without the mechanical stresses that are associated with the traditional manufacturing processes such as mechanical drills,

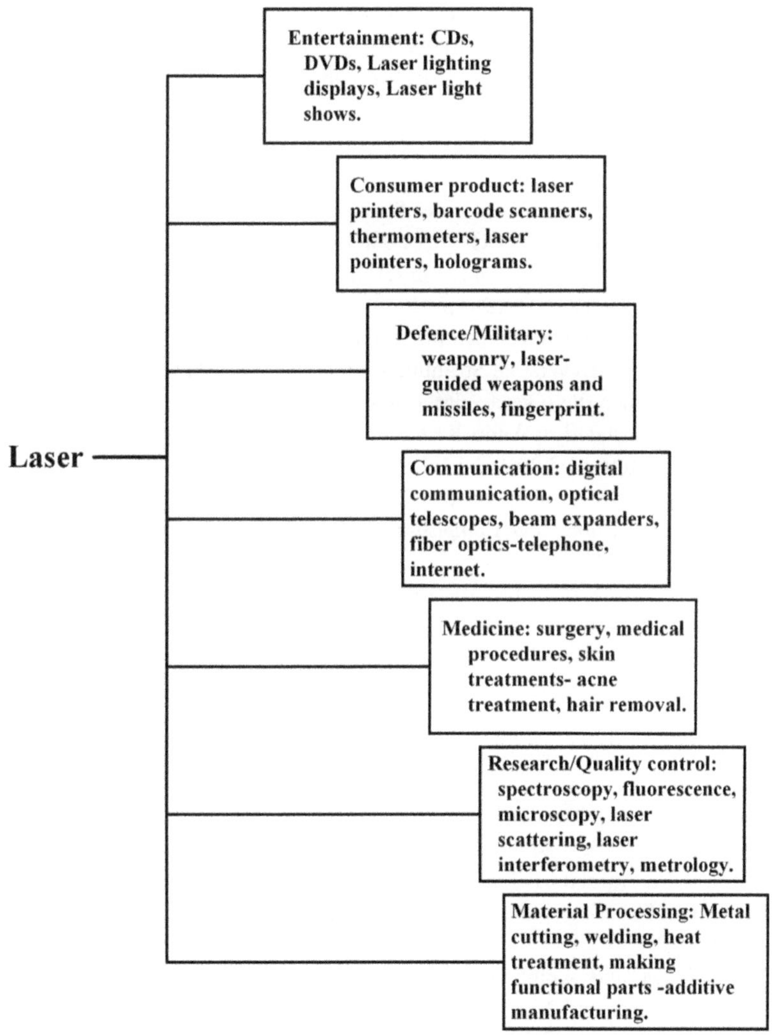

Figure 2.9. A schematic noting some of the applications of lasers [6]. Copyright (2018), with permission of Springer.

which may destroy such miniaturized components. Lasers have made it possible to bring ideas to life and have helped to reduce product to market time through advanced manufacturing processes such as laser additive manufacturing processes, laser cutting as well as laser welding [17–26]. In 2018, Mahamood [6] summarized the application of lasers, as depicted in figure 2.9.

2.4 Conclusion

This chapter presents the working principle of lasers, their properties, different types and areas of application. The working principle and generation of lasers have been described in detail and the properties of lasers were also highlighted. The different types of lasers that are grouped according to the state and type of lasing medium were also presented. Some of the areas of applications of lasers including military, medical and manufacturing industries have been discussed. The role of laser processing in Industry 4.0 is also highlighted in this chapter.

References

[1] Einstein A 1917 Zur Quantentheorie der Strahlung *Phys.* Z **18** 121–8

[2] Planck M 1900 Zur Theorie des Gesetzes der Energieverteilung im Normalspectrum *Verh. Dtsch. Phys. Ges.* 2 237–45 Translated in terHaar D 1967 *The Old Quantum Theory* (PDF) (Pergamon Press) LCCN 66029628, p 82

[3] Planck M 1900 Entropie und Temperatur strahlender Wärme *Ann. Phys. (Berlin)* **306** 719–37

[4] Planck M 1900 Über irreversible Strahlungsvorgänge *Ann. Phys. (Berlin)* **306** 69–122

[5] Planck M 1901 Über das Gesetz der Energieverteilung im Normalspektrum *Ann. Phys. (Berlin)* **4** 553 Translated in Ando K, *On the Law of Distribution of Energy in the Normal Spectrum* (PDF). Retrieved 2019-3-13

[6] Mahamood R M 2018 Laser basics and laser material interactions *Laser Metal Deposition Process of Metals, Alloys, and Composite Materials. Engineering Materials and Processes* (Cham: Springer) 11–35

[7] Stoker M R 2005 Basic principles of lasers *Anaesth. Intensive Care Med.* **6** 402–4

[8] Thomas G and Isaacs R 2011 Basic principles of lasers *Anaesth. Intensive Care Med.* **12** 574–7

[9] Mahamood R M and Akinlabi E T 2017 Influence of scanning speed intermetallic produced *in situ* in laser metal deposited TiC/Ti6Al4V composite *Mater. Technol.* **51** 473–8.

[10] Mahamood R M and Akinlabi E T 2015 Effect of processing parameters on wear resistance property of laser material deposited titanium–alloy composite *J. Optoelectron. Adv. Mater.* **17** 1348–60

[11] Mahamood R M and Akinlabi E T 2015 Laser metal deposition of functionally graded Ti6Al4V/TiC *Mater. Des.* **84** 402–10

[12] Mahamood R M, Akinlabi E T and Owolabi M G 2017 Laser metal deposition process for product remanufacturing *Advanced Manufacturing Technologies* ed K Gupta (Cham: Springer) 267–91

[13] Mahamood R M and Akinlabi E T 2018 Advanced Noncontact Cutting and Joining Technologies: Micro- and Nano-Manufacturing (Cham: Springer)

[14] Akinlabi E T, Mahamood R M and Akinlabi S A 2016 *Advanced Manufacturing Techniques Using Laser Material Processing* (Hershey, PA: IGI Global)

[15] Weber R, Michalowski A, Abdou-Ahmed M, Onuseit V, Rominger V, Kraus M and Graf T 2011 Effects of radial and tangential polarization in laser material processing *Phys. Procedia* **12** 21–30

[16] Otto A and Schmidt M 2010 Towards a universal numerical simulation model for laser material processing *Phys. Procedia* **5** 35–46

[17] Pityana S, Mahamood R M, Akinlabi E T and Shukla M 2013 Gas flow rate and powder flow rate effect on properties of laser metal deposited Ti6Al4V *IMECS 2013 Int. MultiConf. Eng. Comput. Sci. (Hong Kong, 13–15 March 2018)* vol II pp 848–51

[18] Lu L, Qian M, Tang H P, Yan M, Wang J and StJohn D H 2016 Massive transformation in Ti–6Al–4V additively manufactured by selective electron beam melting *Acta Mater.* **104** 303–11

[19] Jain A and Singh B 2020 Parametric analysis during laser cutting of basalt–glass hybrid composite *Lasers Manuf. Mater. Process.* **7** 111–39

[20] El-Batahgy and Kutsuna M 2009 Laser beam welding of AA5052, AA5083, and AA6061 aluminum alloys *Adv. Mater. Sci. Eng.* **974182** 9

[21] Svenungsson J, Choquet I and Kaplan A F H 2015 Laser welding process—a review of keyhole welding modelling *Phys. Procedia* **78** 182–91

[22] Mei L, Chen G, Jin X, Zhang Y and Wu Q 2009 Research on laser welding of high-strength galvanized automobile steel sheets *Opt. Lasers Eng.* **47** 1117–24

[23] Khan M M A, Romoli L, Fiaschi M, Sarri F and Dini G 2010 Experimental investigation on laser beam welding of martensitic stainless steels in a constrained overlap joint configuration *J. Mater. Process. Technol.* **210** 1340–53

[24] Tenner F, Brock C, Klampfl F and Schmidt M 2015 Analysis of the correlation between plasma plume and keyhole behavior in laser metal welding for the modeling of the keyhole geometry *Opt. Lasers Eng.* **64** 32–41

[25] Akbari M, Saedodin S, Toghraie D, Shoja Razavi R and Kowsari F 2014 Experimental and numerical investigation of temperature distribution and melt pool geometry during pulsed laser welding of Ti6Al4V alloy *Opt. Laser Technol.* **59** 52–9

[26] Courtois M, Carin M, Le Masson P, Gaied S and Balabane M 2013 A new approach to compute multi-reflections of laser beam in a keyhole for heat transfer and fluid flow modelling in laser welding *J. Phys. D: Appl. Phys.* **46** 505305

IOP Publishing

Laser Micro- and Nano-Scale Processing
Fundamentals and applications
Ahmed Issa and Dermot Brabazon

Chapter 3

Physical principles of laser–material interaction regimes for laser machining processes

Ahmed Issa, Furat I Hussein Al-Najjar, Ahmed Al-Hamaoy and Bassam G Rasheed

Lasers, with their unique characteristics in terms of excellent beam quality, especially directionality and coherency, make them the solution that is key for many processes that require high precision. Lasers have good susceptibility to integrate with automated systems, which provides high flexibility to reach difficult zones. In addition, as a processing tool, a laser can be considered as a contact-free tool of precise tip that became attractive for high precision machining at the micro and nanoscales for different materials. All of the above advantages may be not enough unless the laser technician/engineer has enough knowledge about the mechanism of interaction between the laser light with the processed material.

Several sequential phenomena occur when an intense laser beam is incident on the surface of a material. Heating, melting, vaporization and plasma formation are present in the normal interaction of an intense laser beam with matter. This may be followed by additional events such as acoustic and optical emissions, structure shockwaves, thermal effects, structural defects and residual stresses. The process is affected by a lot of variables that can transfer the interaction towards extremely different behavior in terms of colder and fewer side-effect interactions, which yield precise features for the processed material. The most crucial variables are the time scale of interaction and laser wavelength with respect to the properties of the processed material undertaken as well as the laser fluence.

The objective of this chapter is to introduce the fundamentals of physical and mathematical concepts of laser and matter interaction and its dependency on different time scale regimes. Interaction with a short and ultra-short laser pulse has attracted a significant amount of interest in industry due to its huge impact in micro-/nanomachining applications.

3.1 Introduction

Basically, in pulsed laser operation, the optical power is dumped out in the form of bursts at a defined duration and repetition rate [1]. A wide range of technologies addressing a number of diverse motivations are utilized to produce various types of pulsed laser apparatus of different time scale outputs. The objective is to fulfill the requirements of numerous applications, especially laser materials processing applications [2, 3].

It is well known that laser micro-/nanomachining processes are based on material ablation as a result of laser–matter interaction processes [4]. The important criteria are the energy absorption and heat transfer mechanisms which depend on laser beam characteristics, process conditions and material properties. Peak power, pulse width and laser wavelength are among the important factors in directing the interaction between laser light and material towards either thermal or photochemical types [5].

Numerous issues affect the interaction mode between the laser light and matter. The time scale issue occupies the first order. The incident electromagnetic radiation on the surface of a material is absorbed by the free electrons of atoms. The latter follow the high frequency of the laser due to their light weight (compared with nuclei) [6]. Depending on the type of material and incident photon energy, photon interaction with matter is related to coupling with electronic or vibrational states in the material. The specific interaction mechanism results in absorption for a fraction of the optical energy. In metals, the inverse Bremsstrahlung effect predominates where photon–electron interactions and the resulting photons' absorption occur in free electrons in a time scale of 10^{-16} s. The energized electrons share their energies with other electrons, in a time scale of 10^{-15} to 10^{-14} s, via electron–electron collisions. This process of optical absorption can be accelerated with increasing the temperature of the electron system. The absorbed energy transfers to the lattice through electron–phonon interaction within the time of excitation in a time scale of about 10^{-11}–10^{-12} s (1–10 ps). As a result, the laser heats up the target material during the pulse duration life, before energy transfers towards the adjacent zones via conduction, as illustrated in figure 3.1 [7–9].

In dielectric materials, such as ceramics and polymers, the bounded electrons do not easily follow the high frequencies of the laser. Photon–phonon interactions

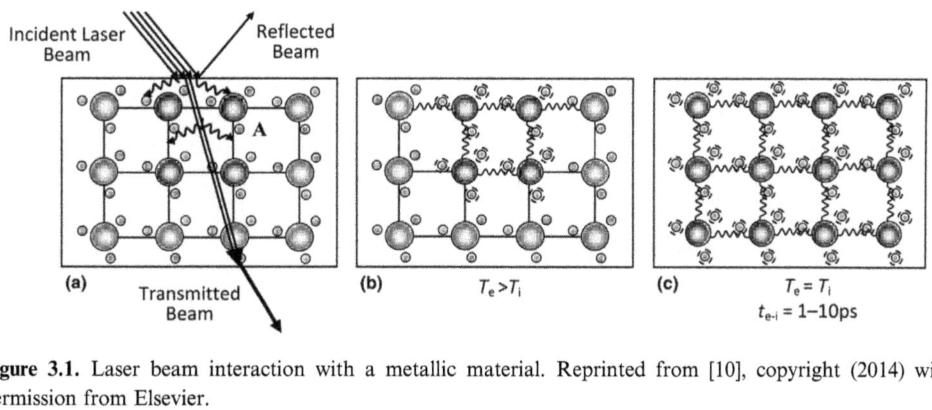

Figure 3.1. Laser beam interaction with a metallic material. Reprinted from [10], copyright (2014) with permission from Elsevier.

predominate, especially when the laser wavelength (or photon energy) more suitably matches the excitation energy of phonon.

In semiconductors, resonant excitations such as transitions of electrons from the valence to conduction bands or within bands is the cause for the absorption of laser light. The band gap energy value is crucial for absorbing the laser radiation with respect to the laser photon energy. When the photon energy is equal or greater than the band gap energy, photon absorption by the valence band electrons through interband transition occurs (i.e. from valence band to conduction band). When electrons occupy the conduction band partially they absorb laser energy in the same mechanism as metals. On the other hand, when the photon energy is less than the band gap energy, absorption may not be achieved and can be disrupted, especially at low intensities. In irradiation with shorter wavelength radiation, absorption increases due to the electrons' excitation [7, 11, 12].

3.2 Time scale role

Continuous development of pulsed laser devices of short pulses, from nanoseconds towards femtoseconds, has effectively contributed to the fabrication of extremely precise edges compared with millisecond and continuous wave (CW) lasers. When a material surface is irradiated with a stream of laser pulses, optical energy conversion into thermal energy occurs in a time duration of 10^{-13} for metals and 10^{-12}–10^{-6} for nonmetals [13]. The mode of interaction between a laser and a material depends on three time scales of pulse duration (τ), electron cooling time (t_e) and lattice heating time (t_l). These three time scales define three regimes of interactions between the laser and materials: nanosecond, picosecond and femtosecond regimes.

The electron cooling time is much less than the lattice heating time ($t_e \ll t_l$) due to less heat capacity of the electron than that of the lattice. These facts define two modes of interactions, thermal (photothermal) and non-thermal (photochemical), between laser and material. When a material is processed with three different pulse durations (nanosecond, picosecond and femtosecond) the precision and quality of the produced structures via ablation will be significantly different. Growing evidence reveals that the solution for fabricating extremely precise microstructures is the ultra-short laser pulses of picosecond and femtosecond time scales. Employing nanosecond or longer laser pulses will contribute to degradation of the requested quality due to thermal effects. It is worth mentioning that sometimes shortening the pulse length does not necessarily yield good quality results. However, shortening pulse duration is not the only factor to attain processing quality. Material type and laser wavelength, also, play important roles in the precision of material ablation [14–16].

3.3 Laser–matter interaction

Understanding the mechanisms of the interaction between an intense laser beam with material is the key for technological applications in laser material processing. When a high-power laser is focused on a solid target, absorption of the incident light occurs within the few atomic layers accompanied with instantaneous and rapid increase in the surface temperature [17]. When irradiating with a pulsed laser, the energy fluence heats up a layer of the surface at the end of the pulse to a depth called the optical penetration depth, also named the skin depth. At that depth, the beam intensity is about $1/e$ of its original value of intensity at the surface. From the reciprocal for the absorption coefficient value, the optical penetration depth (l_{op}) at which the laser beam is attenuated can be expected:

$$l_{\mathrm{op}} = \frac{1}{\alpha} = \frac{\lambda}{4\pi k} \tag{3.1}$$

where α is the absorption coefficient, λ is the laser wavelength and k is the extinction coefficient of the material.

The amount of conducted heat and other related physical phenomena depend on the above-mentioned time scale of the pulses as well as the material time scales of the electrons and lattice. Unlike femtosecond pulses, irradiating with nanosecond pulses causes heat accumulation and flow towards the bulk via conduction to a length called the heat diffusion length (l_{d}):

$$l_{\mathrm{d}} = \sqrt{D\tau} \tag{3.2}$$

where D is the thermal diffusivity and τ is the pulse duration.

Equation (3.2) states the heat diffusion length is a function of both the thermal characteristics of the irradiated material and pulse time length. Above a critical threshold of heat accumulation, a melt pool initiates around which the bulk of the material heats up. During the melt phase time, the adjacent zone to the melting pool undergoes changes in its microstructure and properties and is termed the heat affected zone (HAZ) [18, 19].

Melting can be reached rapidly in about 300 ns with a laser intensity of 10^9 W cm^{-2}. Certainly, increasing the intensity will accelerate melting. In other words, increasing the intensity tenfold will reduce the melting time tenfold too. Increasing the intensity also accelerates the vaporization rate and causes shockwaves inside the bulk with speeds reaching 3 km s^{-1}. Due to superheating at the end of the pulse and the created high-pressure vapor inside the melt, expulsion of the molten material ejected with high-pressure vapor occurs outside the interaction zone. Some of the ejected molten materials, which were not totally vaporized, solidify and are recast around the interaction zone.

The heat exchange between the electrons absorbing the laser radiation and the lattice results in heat flow inside the material. Assuming that the thermal equilibrium within the electronic-lattice subsystems is verified, the process can be expressed by the following one-dimensional, two-temperature heat conduction model equations:

$$C_{\mathrm{e}}\left(\frac{\partial T_{\mathrm{e}}}{\partial t}\right) + C_{\mathrm{l}}\left(\frac{\partial T_{\mathrm{l}}}{\partial t}\right) = \frac{\partial}{\partial z}\left(K_{\mathrm{e}}\frac{\partial T_{\mathrm{e}}}{\partial z}\right) + Q(z, t) \tag{3.3}$$

$$C_{\mathrm{l}}\left(\frac{\partial T_{\mathrm{l}}}{\partial t}\right) = \Gamma_{\mathrm{e-p}}(T_{\mathrm{e}} - T_{\mathrm{l}}) \tag{3.4}$$

The applied laser source can be represented as a Gaussian heat source:

$$Q(z, t) = \sqrt{\frac{\beta}{\pi}}\frac{F(1 - R)}{\tau}\alpha \exp\left(-\alpha z - \beta\left(\frac{t - 2\tau}{\tau}\right)^2\right) \tag{3.5}$$

where C_e and C_l are the heat capacities for the electron and lattice per unit volume, respectively, T_e and T_l are the electron and lattice temperatures, respectively, β is a constant equal to 4 ln 2, F is the incident laser fluence, K is the thermal conductivity of the electrons, z is the coordinate perpendicular to the surface of the target, $Q(z,t)$ is the supplied heat flux along the z-axis, and Γ_{e-p} is the electron–phonon coupling strength constant.

The two-temperature model of equation (3.3) has considered the following:

The thermal conductivity (K_l) for the lattice (phonon component) is neglected.

The evolution of the electron and lattice temperatures during heat flux supply with $Q = 0$.

The electron–phonon coupling constant Γ_{e-p} is a parameter for characterizing the energy exchange between the electrons and the lattice.

The processed material is homogeneous.

The macroscopic properties, heat capacity and thermal conductivity are temperature dependent.

The electrons and lattice are in broken equilibrium between them but each system has local equilibrium with itself.

Three time scales τ, t_e and t_p, where τ is the laser pulse duration, t_e is the electron cooling time ($t_e = C_e/\Gamma_{e-p}$) and t_l is the lattice heating time ($t_l = C_l/\Gamma_{e-p}$). The electron cooling time is much less than the lattice heating time ($t_e \ll t_l$). It is useful to model these evolution times.

The incident laser energy on the surface heats up the electrons to very high transient temperatures. This is related to the fact that the heat capacity of electrons is much less than the heat capacity of the lattice. As the electron temperature (in units of energy) is still less than the Fermi energy, the heat capacity of the electrons and nonequilibrium electron thermal conductivity are given by:

$$C_e = C_e' T_e \tag{3.6}$$

and

$$K_e = K_o(T_l) \cdot \frac{T_e}{T_l} \tag{3.7}$$

where C'_e is a constant and $K_o(T_l)$ is the conventional equilibrium thermal conductivity of a metal.

Irradiation with a laser beam for a material surface, in a perpendicular direction, for a certain time (t) will heat up the target and raise its temperature according to:

$$\Delta T(x, t) = 2(1 - R)\alpha I_o \frac{t}{\pi k \rho C} \text{ierfc} \frac{z}{\sqrt{\frac{Kt}{\rho C}}} \tag{3.8}$$

where I_o is the spatial distribution of laser intensity and ρ is the density of the materials.

When the depth z is greater than $\sqrt[4]{Kt/\rho C}$, equation (3.8) can be reduced to [20–23]:

$$\Delta T(t) = \frac{2\alpha I_o \sqrt{t}}{\sqrt{\pi k \rho C}}. \tag{3.9}$$

3.4 Interaction regimes in laser–matter interaction

The interaction regime type between a laser and material mainly depends on the three time scales mentioned in section 3.2. These parameters define the three interactions of nanosecond, picosecond or femtosecond regimes as well as the resulting phenomena and produced features. The associated physical processes depend on time scales of material regarding the incident pulse width of efficient intensity irradiation. Figure 3.2 illustrates the physical processes that occur when an intense single pulse is incident on a target material. Nonlinear processes such as multi-photon absorption and/or tunneling ionization can occur after one femtosecond. Avalanche ionization occurs after 50 fs. Thermalization, i.e. reaching thermal equilibrium, of the electrons of the target material takes place after 100 fs. Other processes can appear with wider pulses, after one picosecond the energy transforms from the electrons towards the lattice. Thermodynamic processes take place after 10 ps such as thermal diffusion, fusion and explosion. With wider pulses of more than 1 ns, photochemical processes such as phase transformation occur [1, 24].

3.4.1 Thermal interaction (when $\tau > 1$ ns $\gg t_l \gg t_e$)

Such mode of interaction is achieved when the pulse duration is greater than t_e and t_l time scales. The range of milliseconds towards the nanosecond pulse duration

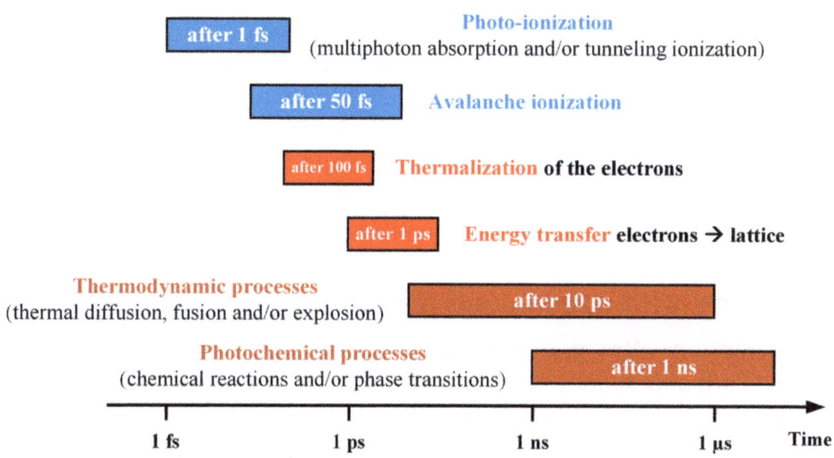

Figure 3.2. Occurrence of different physical phenomena during the different time scales involved in laser–material interaction, modified based on. Reprinted from [24], copyright Walter de Gruyter GmbH.

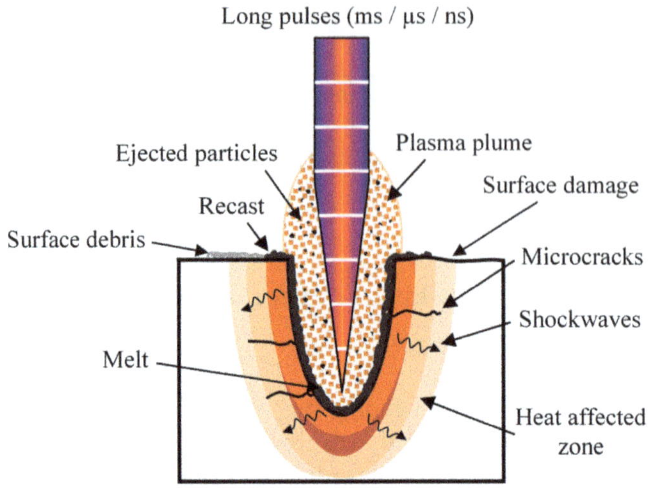

Figure 3.3. Interaction of a long pulse and CW with a material, modified based on. Reprinted from [10], copyright (2014) with permission from Elsevier.

ranges, or infinite range for CW mode of operation, are utilized for these kinds of interactions. In this case there will be enough time for reaching electron–lattice thermal equilibrium and heat conduction towards the bulk. As a result, heating, melting, vaporization and photoionization from the liquid state are the fundamental processes in this time scale. Usually, these processes are followed by expansion and thermal stresses in the bulk of the processed material [25]. Such mode of interaction shows a significant generated HAZ with a width directly proportional to the pulse width length as depicted in figure 3.3 [10].

3.4.2 Non-thermal interaction ($\tau \ll t_l \ll t_e$)

In this mode the pulse duration is shorter than t_e and t_l time scales such as the femtosecond or picosecond pulse durations. Electrons are heated instantly and transfer their energy to their lattices in about 1 ps. If the laser's ultra-short pulse duration with high enough pulse energy conditions are met, breaking of lattice bonding instantly occurs without having time to transfer the energy to the neighboring lattice. As a result, a direct solid–vapor transition occurs without heat conduction into the target as well as a great reduction of the HAZ as illustrated in figure 3.4. This type of interaction is also called photochemical interaction [10, 26].

3.5 Processing with nanosecond pulses

With the presence of ultra-short pulses, the nanosecond pulses are considered long pulses when compared with picosecond and femtosecond lasers. Pulse duration is the time at which the material is excited with coherent radiation. The mechanism of material removal via the vaporization phase is called ablation [27].

Many research works reveal the direct relationship between the pulse duration with the ablation threshold and heat penetration depth [28]. Increasing the pulse width

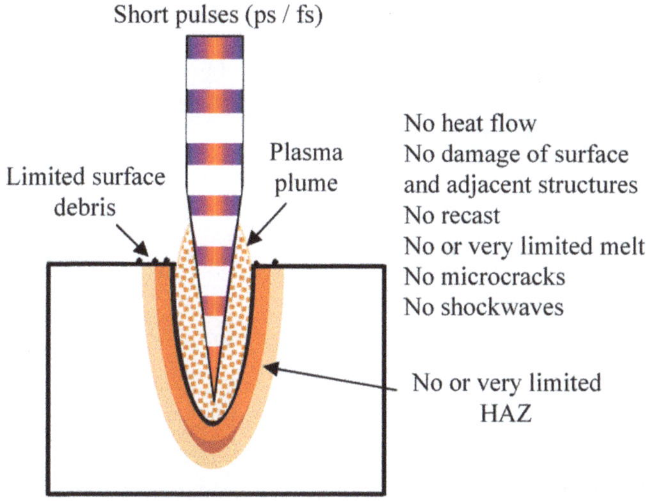

Figure 3.4. Ultra-short pulses interaction with a material, modified based on. Reprinted from [10], copyright (2014) with permission from Elsevier.

will increase the pulse energy too and consequently the fluence (energy density). In nanosecond laser machining, material removal passes through heating, melting and vaporization phases. Thus, an important percentage of material removal via ablation passes through solid–liquid–vapor phases. Small amounts of removed and ejected molten materials cling at the edges of the machined features and then re-solidify. In addition, the ejected molten material splashes around the processed zone and contributes to the degradation of precision and quality of machining as seen in the comparison presented in figure 3.5 [10, 27].

Lengthening the pulse duration lowers the intensity leading to more penetration of heat towards the bulk of the material, through conduction, resulting in a wider ablated area with less depth. The value of energy required for vaporization of the material is higher than that for achieving melting. At low laser intensities, the generated vapor is transparent relative to the incident radiation. At first approximation, due to the large width of nanosecond pulses with respect to electron and lattice heating times ($\tau \gg t_p$), the electron and lattice temperatures are equal ($T_e = T_p = T$). Thus, thermal conduction inside the target material is considered and the electron–lattice system reaches thermal equilibrium and the heat equation (3.3) can be modified as [21]:

$$C_l\left(\frac{\partial T}{\partial t}\right) = \frac{\partial}{\partial z}\left(K_o\frac{\partial T}{\partial z}\right) + Q(z) \tag{3.10}$$

The depth of penetrated heat (l_{th}) via conduction inside the material is given by:

$$l_{th} = \sqrt{\frac{K\tau}{\rho C_l}} \tag{3.11}$$

where D is the thermal diffusivity, τ is the pulse duration, K is the thermal conductivity, ρ is the target density and C_l the specific heat of the material.

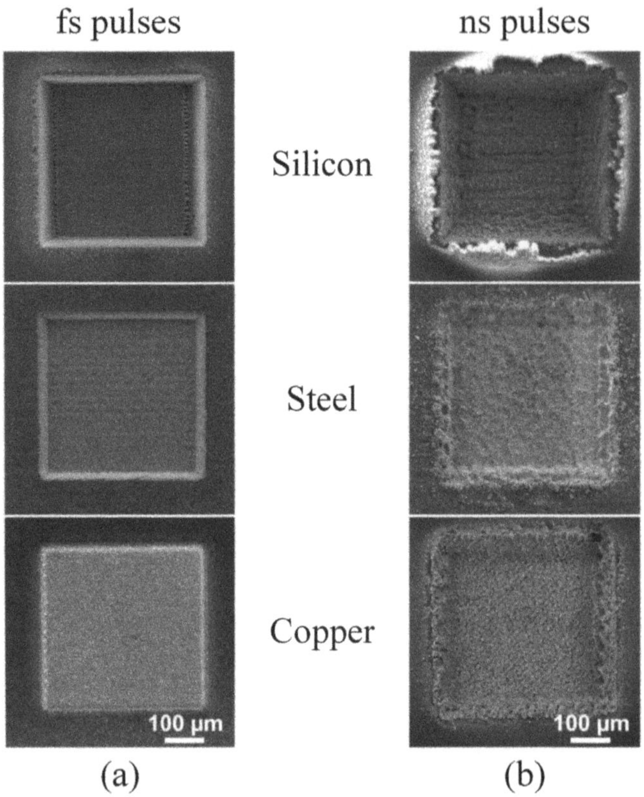

Figure 3.5. Processed features with: (a) femtosecond laser pulses and (b) nanosecond laser pulses, modified based on. Reprinted from [29], copyright (2020) with permission from Elsevier.

Usually the condition $D\tau\alpha^2 \gg 1$ is fulfilled when materials are processed through laser ablation with long pulses. In this case the depth of penetrated heat is much larger than the optical depth ($l_{\text{th}} \gg l_{\text{op}}$). The energy per unit mass (E_{m}) that is deposited inside the target is given by:

$$E_{\text{m}} = \frac{I_{\text{a}}t}{\rho l_{\text{th}}} \tag{3.12}$$

where I_{a} is the absorbed intensity.

Considerable evaporation occurs after a specified time (t_{th}) when the deposited energy becomes larger than the specific heat of evaporation Ω (per unit mass). Powerful vaporization requires the condition $E_{\text{m}} > \Omega$ (or $\tau > t_{\text{th}}$) to be fulfilled, which can be expressed in terms of required effective laser intensity or fluence that exceeds the threshold intensity (I_{th}) or threshold fluence (F_{th}):

$$I > I_{\text{th}\sim} \frac{\rho\Omega\sqrt{D}}{\sqrt{\tau}} \tag{3.13}$$

$$F > F_{\text{th}\sim}\rho\Omega\sqrt{D\tau} \tag{3.14}$$

where the threshold fluence for vaporization per unit mass is given by ($F_{\text{th}} \approx \rho\Omega/\alpha$). It can be seen from equation (3.14) that the threshold fluence for evaporation increases with long pulses [21].

After a period of time, nanosecond pulses heat up the surface of the target to the melting point and then are followed by a vaporization point. Ablation occurs due to the sequence of these three phase changes. The ablation depth (Z_n) per pulse can be expressed as:

$$Z_n \approx \sqrt{l_{th}}\, Ln\sqrt{\frac{F_a}{F_{th}}} \tag{3.15}$$

Previous equations show that the absorbed fluence has a direct relationship with the pulse width, which in turn has the same relationship with the ablation depth [18].

3.6 Processing with picosecond pulses

In laser ablation of material with picosecond laser pulses, the pulse width is considered in a range greater than the electron cooling time t_e and less than the lattice heating time t_p ($t_e \ll \tau \ll t_p$). The free electrons are responsible for the large amount of heat flow. For a time much greater than the electron cooling time ($t \gg t_e$), which is equivalent to $C_e\, T_e/t \ll \Gamma_{e\text{–}p}\, T_e$, equation (3.3) for the electron temperature becomes quasi-stationary and recast to:

$$C_l\left(\frac{\partial T_l}{\partial t}\right) = \frac{\partial}{\partial z}\left(K_e\frac{\partial T_e}{\partial z}\right) + Q(z) \tag{3.16}$$

The lattice temperature for a given time can be represented by

$$T_p = T_o + \frac{1}{t_p}\int_0^t\left(\frac{-t-\theta}{t_p}\right)T_e(\theta)d\theta \tag{3.17}$$

where θ is the time duration for the given laser pulses.

By neglecting the initial temperature T_o and considering for a time t being much less than the lattice heating time ($t \ll t_p$), equation (3.17) can be reduced to a more simplified form because of the quasi-stationary character of the electron temperature:

$$T_p = T_e\left(1 - \exp\left(\frac{-t}{t_p}\right)\right) \approx \left(\frac{-t}{t_p}\right)T_e \tag{3.18}$$

According to equation (3.18), for nanosecond ablation processes, the lattice temperature can be neglected due its much lower value when compared with electron temperature. When the condition $K_e\, T_e\, \alpha^2 \ll \Gamma_{e\text{–}p}\, T_e$, the analysis of equations (3.16)–(3.18) becomes simpler. During interaction, heat exchange between the heated electrons with the lattice is done, at the end of the pulse time the electron and lattice temperatures are given by:

$$T_e = \frac{I_a\alpha}{\Gamma_{e-p}}\exp(-\alpha z) \quad (3.19)$$

$$T_p = \frac{F_a\alpha}{C_i}\exp(-\alpha z) \quad (3.20)$$

In picosecond laser ablation, the heat conduction is the minor part where the major part of heat is consumed in local interaction. Picosecond pulses perform very well with micromachining processes because of the direct phase change from solid to vapor (or plasma). However, this doesn't prevent very small traces of liquid phase formation inside the interaction zone and the edges. The ablation depth per pulse can be presented by:

$$Zn \approx \alpha^{-1}Ln\sqrt{\frac{F_a}{F_{th}}} \quad (3.21)$$

Comparing with nanosecond laser machining, the unique property of picosecond laser machining is that it delivers focused energy at strongly localized areas. This results in very limited HAZ in the surrounding material. In other words, less amount of melt initiates, which yields higher accuracy [21, 30].

3.7 Processing with femtosecond pulses

When processing materials with femtosecond pulses, the pulse duration is much less than the electron cooling time ($\tau \ll t_e$). This is considered an extremely short time scale of interaction between the laser and material. The material removal by ablation passes through a direct solid–vapor (or solid plasma) transition. Here, the time of lattice heating is on the picosecond time scale. The concentration of relatively huge energy inside a little volume of matter yields vapor creation and plasma formation, which is followed by outside expansion without noticeable heat conduction towards the bulk of the material. The created features from such an interaction have precise and fine surfaces and edges [31, 32].

With such a period of time, the electron–lattice coupling is ignored and equation (3.3) can be solved in a more simplified manner. In addition, for a very short time of interaction of femtosecond pulse duration where the condition $D_e\tau < \alpha^2$ is fulfilled, the conduction term is neglected in the same equation. In that case, the depth of heat penetration is less than the optical depth ($l_{th} < l_{op}$). Accordingly, equation (3.3) can be recast as:

$$C_e'\frac{\partial T_e^2}{\partial t} = 2\, I_a\alpha \exp(-\alpha z) \quad (3.22)$$

This yields the electron temperature at a specified time $T_e(t)$:

$$T_e(t) = \sqrt{(T_0^2 + \frac{2\, I_a\alpha}{C_e'}t \exp(-\alpha z)} \quad (3.23)$$

where T_o is the initial temperature ($T_o = T_e(0)$) and I_a is the absorbed intensity ($I_a = I_o(1 - R)$).

Assuming the initial conditions, the initial temperature is equal to the electron temperature and lattice temperature ($T_e = T_l = T_o$). At the end of the laser pulse time where the interaction of the pulse with the target vanishes, by knowing the absorbed laser fluence (F_a), the temperature of the target is expressed by:

$$T_e(z, \tau) \approx \sqrt{\frac{\alpha\,(1 - R)\,F_\alpha}{\rho C}}\,\exp\!\left(-\frac{\alpha z}{2}\right) \tag{3.24}$$

$$F_a = I_a \tau \tag{3.25}$$

Even in the femtosecond range, thermal ablation processes occur and reduce the quality, accuracy and efficiency of micromachining [18]. The transferred heat towards the lattice and then to the bulk of the material through conduction causes transient electrons cooling at the end of the pulse in a very short time.

Due to the energy transfer to the lattice and heat conduction of the bulk material, the electrons are rapidly cooled after the laser pulse:

$$T_l \approx T_e^2\!\left(-\frac{\alpha z}{2}\right) \tag{3.26}$$

The attainable lattice temperature is determined by the average cooling time of the electrons ($t_e{}^a$):

$$t_e^a \approx \frac{C_e' T_e(\tau)}{2\,\Gamma_{e-p}} \tag{3.27}$$

$$T_l \approx \frac{F_a \alpha}{C_l}\exp(-\alpha z) \tag{3.28}$$

Equation (3.28) can be written for conducting the fluence required to attain strong evaporation if the condition of ($C_l T_l > \rho\ \Omega$) is fulfilled:

$$F_a \geqslant F_{th}\exp(\alpha z) \tag{3.29}$$

In femtosecond processing via ablation, the major part of the absorbed energy is stored in a thin layer. Here, a very limited, neglected, part is transferred to the lattice. The depth of removal per pulse via ablation is given as [18, 21, 30]:

$$Z_n \approx \alpha^{-1} Ln\sqrt{\frac{F_a}{F_{th}}} \tag{3.30}$$

Based on the two-temperature equation, the mass of ablated material from the interaction zone can be estimated as follows:

$$m = N\rho A\alpha^{-1} Ln\frac{F_a}{F_{th}} \tag{3.31}$$

where m is the ablated material mass, N is the number of irradiated pulses, ρ is the density and A is the focal spot area.

3.8 Summary

This chapter introduced the physical principles that govern laser–material interaction phenomena exhibited during some famous laser machining processes. Energy absorption mechanisms are explained with relevance to timescale effects upon laser irradiation of metallic and dielectric materials. Furthermore, the chapter introduces several mathematical and heat conduction models that measure the introduced effects due to laser irradiation of materials. The chapter also discusses the interaction regimes due to different laser pulse durations ranging from short to ultra-short pulses. The overall chapter aim is to introduce a base-understanding for researchers on what to expect upon irradiating materials with laser beams. This understanding is delivered in terms of the laser–material interaction regimes and the most important parameters that govern these regimes leading to understanding which governs: accuracy of micro- and nano-machining, minimized HAZ, how ablation or defects are induced, ablation thresholds, etc.

References

[1] Royon A, Petit Y, Papon G, Richardson M and Canioni L 2011 Femtosecond laser induced photochemistry in materials tailored with photosensitive agents *Opt. Mater. Express* **1** 866–82

[2] Brown M S and Arnold C B 2010 Fundamentals of laser–material interaction and application to multiscale surface modification(Springer Series in Materials Science vol 135) ed K Sugioka, M Meunier and A Piqué (Berlin, Heidelberg: Springer) pp 91–120

[3] Hecht J 2019 *Understanding Lasers* 4th edn (Hoboken, NJ: Wiley)

[4] Ahmmed K M T, Grambow C and Kietzig A-M 2014 Fabrication of micro/nano structures on metals by femtosecond laser micromachining *Micromachines* **5** 1219–53

[5] Jelínková H 2013 *Lasers for Medical Applications; Diagnostics, Therapy and Surgery* (Philadelphia, PA: Woodhead)

[6] Dahotre N B and Harimkar S P 2008 *Laser Fabrication and Machining of Materials* 2nd edn ed N B Dahotre and S P Harimkar (New York: Springer)

[7] Migliore L 1996 *Laser Materials Processing* (New York: Marcel Dekker)

[8] Yadava S M V 2015 Laser beam micromachining (LBMM)—a review *Opt. Lasers Eng.* **73** 89–122

[9] Zhang J, Tao S, Wang B and Zhao J 2016 Studies on laser material processing with nanosecond and subnanosecond and picosecond and sub-picosecond pulses, *Laser Applications in Microelectronic and Optoelectronic Manufacturing (LAMOM) XXI (San Francisco, CA, 13–18 Feb 2016) Proc. SPIE* vol 9735 pp 192–203

[10] Nath A K 2014 Laser drilling of metallic and nonmetallic substrates *Reference Module in Materials Science and Materials Engineering* vol 9 (New York: Elsevier) pp 115–75

[11] Hummel R E 2004 *Understanding Materials Science* 2nd edn (New York: Springer)

[12] Majumdar J D and Manna I 2003 Laser processing of materials *Sadhana* **28** 495–562

[13] Rihakova L and Chmelickova H 2015 Laser micromachining of glass, silicon, and ceramics *Adv. Mater. Sci. Eng.* **2015** 584952

[14] Harzic R L, Breitling D, Weikert M, Sommer S, Fohl C, Valette S, Donnet C, Audouard E and Dausinger F 2005 Pulse width and energy influence on laser micromachining of metals in a range of 100 fs to 5 ps *Appl. Surf. Sci.* **249** 322–31
[15] Chichkov B N, Momma C, Nolte S, von Alvensleben F and Tünnermann A 1996 Femtosecond, picosecond and nanosecond laser ablation of solids *Appl. Phys.* A **63** 109–15
[16] Yao Y L, Chen H and Zhang W 2005 Time scale effects in laser material removal: a review *Int. J. Adv. Manuf. Technol.* **26** 598–608
[17] Al-Najjar F I H 2018 Numerical analysis of the effect of scanning speed on the temperature field distribution for laser heat treatment applications *Al-Nahrain J. Eng. Sci.* **21** 213–22
[18] Jackson M J 2006 *Microfabrication and Nanomanufacturing* (Boca Raton, FL: Taylor & Francis)
[19] Fox M 2010 *Optical Properties of Solids* (Oxford Master Series in Physics) 2nd edn (Oxford: Oxford University Press)
[20] Han J and Li Y 2011 Interaction between pulsed laser and materials *Lasers: Applications in Science and Industry* ed K Jakubczak (Rijeka: InTechOpen) pp 109–30
[21] Stafe M, Marcu A and Puscas N N 2014 *Pulsed Laser Ablation of Solids: Basics, Theory and Applications* (Springer Series in Materials Science vol 130) (Berlin: Springer)
[22] Hao-Feng H, Yang J, Yang H, Xiao-Yan D, Xian-Wen L, Jing-Hui G, Xiao-Lei W and Hong-Chen Z 2011 Thermal analysis of intense femtosecond laser ablation of aluminum *Chin. Phys.* B **20** 044204
[23] Miotello A and Ossi P 2010 *Laser–Surface Interactions for New Materials Production* (Berlin, Heidelberg: Springer)
[24] Petit Y *et al* 2018 On the femtosecond laser-induced photochemistry in silver-containing oxide glasses: mechanisms, related optical and physico-chemical properties, and technological applications *Adv. Opt. Technol.* **7** 291–309
[25] Sands D 2011 Pulsed laser heating and melting *Heat Transfer: Engineering Applications* ed V Vikhrenko (Rijeka: InTechOpen) pp 47–70
[26] Chowdhury I H and Xu X 2003 Heat transfer in femtosecond laser processing of metal *Numerical Heat Transfer, Part A* **44** 219–32
[27] Phillips K C, Gandhi H H, Mazur E and Sundaram S K 2015 Ultrafast laser processing of materials: a review *Adv. Opt. Photonics* **7** 684–712
[28] Schoonderbeek A, Biesheuvel C A, Hofstra R M, Boller K-J and Meijer J 2004 The influence of the pulse length on the drilling of metals with an excimer laser *J. Laser Appl.* **16** 85–91
[29] Domke M, Matylitsky V and Stroj S 2020 Surface ablation efficiency and quality of fs lasers in single-pulse mode, fs lasers in burst mode, and ns lasers *Appl. Surf. Sci.* **505** 144594
[30] Hamad A H 2016 Effects of different laser pulse regimes (nanosecond, picosecond and femtosecond) on the ablation of materials for production of nanoparticles in liquid solution *High Energy and Short Pulse Lasers* ed R Viskup (Rijeka: IntechOpen) pp 305–25
[31] Momma C, Chichkov B N, Nolte S, Alvensleben F. v., Tiinnermann A, Welling H and Wellegehausen B 1996 Short-pulse laser ablation of solid targets *Opt. Commun.* **129** 134–42
[32] Gamaly E G, Rode A V, Luther-Davies B and Tikhonchuk V T 2002 Ablation of solids by femtosecond lasers: ablation mechanism and ablation thresholds for metals and dielectrics *Phys. Plasmas* **9** 949–57

Chapter 4

Effective working parameters of laser micro-/ nano-machining

Furat I Hussein Al-Najjar, Ahmed Al-Hamaoy, Bassam G Rasheed and Ahmed Issa

Modern emerged technologies impose development and fabrication of miniaturized parts and devices in the micro- and nano-scale. Producing micro- and nano-featured structures requires nonconventional machining processes where conventional machining processes such as grinding, milling and eroding have failed. New emerging processes, such laser machining processes, are still fraught with almost invincible processes. Micro-/nano-machining are the processes of producing parts, microsystems or features at a scale of a few micrometers and less than one hundred nanometers, respectively.

Precise cutting and clean material removal accompanied with a negligible heat affected zone (HAZ), which are usually the characteristics of laser ablation, have opened a wide door for the evolution of remarkable technologies. This has been demonstrated by applications in different fields such as medicine, biotechnology, materials processing, microelectromechanical systems, electronics and communications. The continuous development in laser technology in terms of ultra-short pulse width, short wavelength and optics technologies has reduced the drawbacks of diffraction-limited processing accuracies. Laser micro-/nano-machining requires the attainment of high fluence and short interaction time to achieve ablation processes in nanofabrication and structuring of different materials. To conduct the optimum desired machining process, it is important to integrally consider a number of laser beam and working parameters. Laser wavelength, beam mode, minimum attainable spot size, peak power, pulse duration, pulse repetition rate and scanning speed are some of the important considerations. Manipulating those parameters is crucial for ideal laser ablation represented by yielding the highest resolution of machining with the least lateral dimensions, acceptable depth and minimal or no melt at the edges. The assembly of laser beam delivery and focusing system with an automation system are the essential factors for workpiece positioning and obtaining the desired dimensions. The objective of this chapter is to review the effective parameters

doi:10.1088/978-0-7503-1683-5ch4

associated with laser machining processes that affect the dimensions and quality of laser machining at the micro-/nano-scales in a simple presentation. The review is supported by demonstrating laser processing techniques applied in the field of micro-/nano-machining such as mask, interferometric and scribing techniques.

4.1 Introduction to micro-/nano-machining process parameters

Studying laser material processing applications is complicated due to the interdisciplinary factors that govern the interaction between the laser beam and materials. Optical energy absorption is controlled by a combination of machining parameters including process parameters as classified in figure 4.1. These different parameters are categorized into three main areas; laser parameters, material properties and the process parameters are the main considerations in laser material processing applications. In general, when laser pulses are incident on a surface, the absorbed energy results in various effects namely heating, melting, ionization, vaporization and plasma plume formation [1, 2]. This order could be altered depending on the working parameters and material properties for the processed target. For example, more intense and shorter pulses alter this sequence towards material removal by instant ionization and vaporization of the material into a plasma plume. This yields more suitable processes based on ablation such as machining and cleaning [3, 4]. In laser micro-/nano-machining all of the employed parameters should be combined to produce, as much as possible, a product close to the ideal quality.

Essentially, laser machining processes are based on focusing short, high-peak power pulses of a laser on a confined zone to achieve material removal via ablation. The shape, feature resolution, edge quality, minimal damage and ultimately product functionality are the essential goals to be obtained in the machined products. These characteristics are highly correlated with the spatial and temporal manner in which the laser energy is applied [14]. Exact harmonization between the process parameters is essential to attain high-precision machining [15]. The effective parameters such as

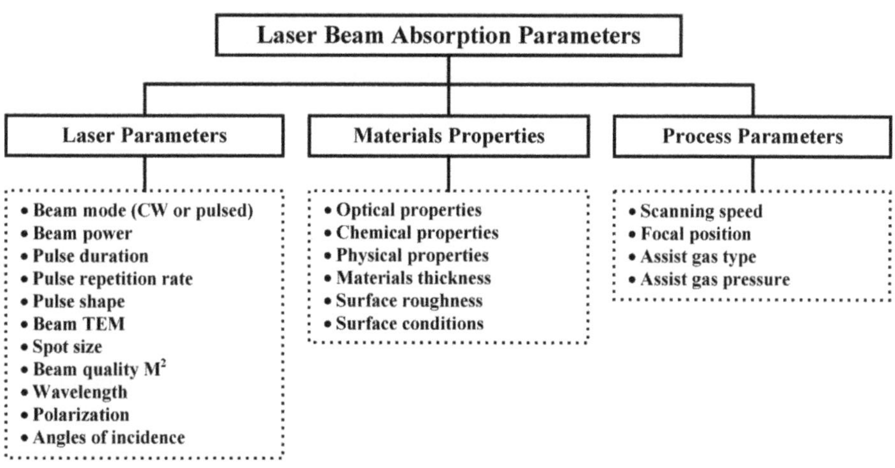

Figure 4.1. The classification of laser beam absorption parameters, depicted from [5–13].

pulse duration, pulse repetition rate, wavelength and scanning speed must be well synchronized to ensure a precise control for the workpiece and avoid side effects of the HAZ [13, 16, 17]. Machining of contours in complex and irregular shapes may cause more energy accumulation at the corners and curves due to the relative deceleration of the laser beam and increase exposure time to laser energy at such zones. Power or machining speed ramping are the solutions used to overcome these problems of overshooting and heating at curved paths and corners [18]. Modern control systems, according to a program, can compensate for some of these parameters such as laser beam power and scanning speed to avoid the problems of curved and corner zones [19].

4.2 Laser beam wavelength

The laser beam wavelength has a significant effect on the type and interaction level of laser light with the material. Many other parameters have a relationship and/or depend on this important factor such as absorption (or optical properties), mode of interaction, spot size and depth of penetration. Due to its importance, many researchers have studied these relationships and effects on micromachining process outcomes [17, 20–22]. This section illustrates the role and effect of laser beam wavelength on the above-mentioned processing parameters.

4.2.1 Wavelength effect on absorption

When laser radiation is incident on a material surface, various phenomena occur including reflection, refraction, absorption, scattering and transmission. The pre-dominating phenomena depend on material type and laser light parameters, wave-length and beam intensity. Absorption of light by the material is related to its complex refractive index (n^*):

$$n^* = n - ik \tag{4.1}$$

where n is the refractive index and k is the extinction coefficient.

The absorptivity (A) and reflectivity (R) of the incident radiation are expressed by:

$$A = 1 - R \tag{4.2}$$

$$R = \frac{\{(n-1)^2 + k^2\}}{\{(n+1)^2 + k^2\}} \tag{4.3}$$

$$A = \frac{4n}{\{(n+1)^2 + k^2\}} \tag{4.4}$$

The coefficients n and k are wavelength- and temperature-dependent and consequently, the reflectivity and absorptivity of the material are significantly affected by these two parameters as well. In general, the absorbance of different materials can be increased at shorter wavelength radiations [23] as illustrated in figure 4.2.

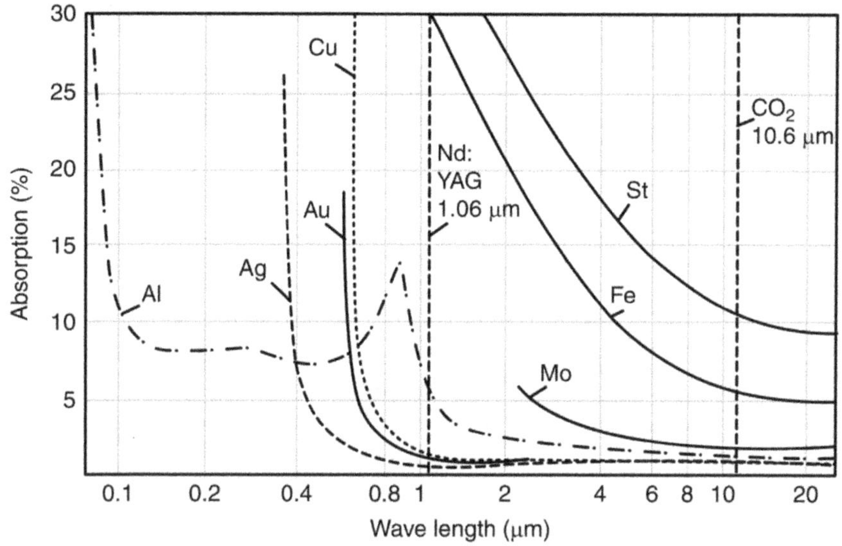

Figure 4.2. Variation of absorbance with wavelength for different metals with highlights on 1.06 μm and 10.6 μm lasers. Reprinted from [23], copyright (2010), with permission from Elsevier.

Depending on the laser type, i.e. wavelength, materials of absorption coefficients in the range of 10^5–10^6 cm^{-1} (optical depth of 10^{-5}–10^{-6} cm) are considered strongly absorbing materials. The interaction of materials with incident light depends on the coefficients n and k. For example, in most metals k is larger than n, therefore 90% of the incident light is reflected at the surface and attenuated to few tens of nanometers beneath the surface. With shorter wavelength radiation, the absorptivity increases at the expense of reflectivity. In general, for nonmetallic materials such as polymers, ceramics and glasses this is reversed as can be seen in figure 4.3. Therefore, the dependency of laser ablation on the absorbance of the incident beam imposes the necessity to have compatibility between the laser wavelength with the requested process [24].

Increasing the temperature of the target has an important role in altering the reflectivity and enhancing the absorptivity. Below the melting points of metals, a shorter wavelength such as that of Nd:YAG lasers is more effective than a longer wavelength of CO_2 lasers, as can be seen in figure 4.4. Once the material melts, the absorptivity increases dramatically, and essentially becomes independent of the wavelength.

4.2.2 Wavelength effect on the interaction mode

In addition to the absorptivity relation with wavelength, the latter affects the mode of interaction with the target as a result of the interaction dependency on the incident photon energy:

$$E_p = h\nu = \frac{hc}{\lambda} \tag{4.5}$$

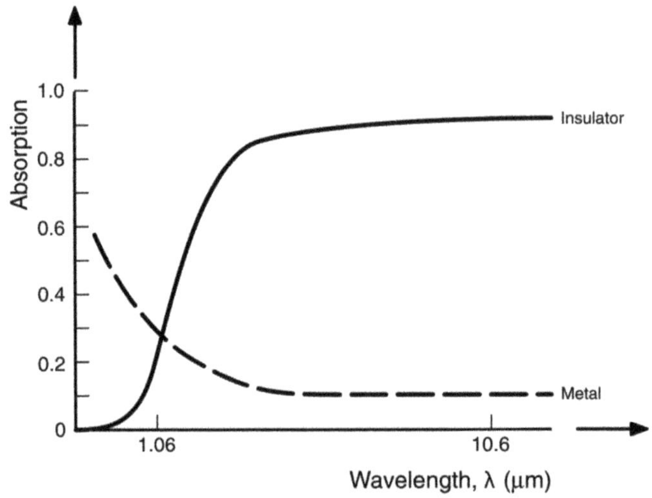

Figure 4.3. Metals tend to absorb short wavelength radiation, on the contrary, dielectric materials absorb long wavelength radiation [25]. John Wiley & Sons, copyright (2009).

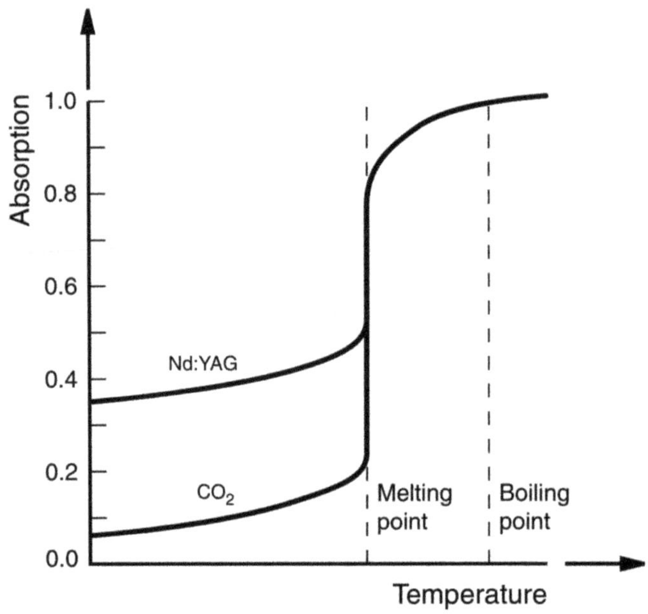

Figure 4.4. The dramatic increase in absorptivity occurs after melting [25]. John Wiley & Sons, copyright (2009).

where E_p is the photon energy, h is Planck's constant (6.626×10^{-34} m^2 kg s^{-1}), v is the photon frequency, c is the speed of light in vacuum (3×10^8 m s^{-1}) and λ is the wavelength.

In some ablation processes, when the energy of the photon equals or exceeds the chemical bond energy, breakdown of the chemical bonds occurs without heat dissipation towards the bulk of material [26]. This process requires a whole avalanche of photons striking the target at the same time to ensure chemical bonds breaking and to avoid instant reformation after the irradiation vanishes [2]. For example, irrespective of other parameters, the KrF excimer laser of 249 nm wavelength has a photon energy of 4.9 eV, which would be sufficient to induce a photochemical ablation process for a single carbon bond of 3.62 eV energy. In comparison, a Nd: YAG laser of 1064 nm, whose photon energy is 1.16 eV, would not be suitable for such a process. It is worth mentioning that photochemical interaction with short wavelength pulses has the same effect with femtosecond, picosecond and even some nanosecond pulses. The combination of short wavelength and ultra-short pulses for a laser beam would result in extreme photon saturation more easily.

4.2.3 Wavelength effect on spot size and depth of penetration

When a Gaussian beam is focused by a lens, a minimum spot size in the order of the wavelength can be attained. Figure 4.5 shows a Gaussian laser beam focused with a focusing lens. If the distance from the laser to the lens is small, the beam divergence is little so that the beam diameter has not increased much. By neglecting the effects of lens aberration, the minimum spot size is given by:

$$\omega = f\theta \tag{4.6}$$

where ω the minimum spot size, f is the focal length of the lens and θ is the beam divergence angle in radiance.

The diffraction-limited beam divergence angle θ is approximated as ($\theta = \lambda/D$), and by assuming the beam fills the aperture of the lens then the minimum spot size can be rewritten as:

$$\omega = f\frac{\lambda}{D} = \lambda \cdot F\# \tag{4.7}$$

where D is the diameter of the limiting aperture and $F\#$ is the F-number of the lens.

The F-number is defined as a measure for the ability of a lens to gather light. It is the ratio of the focal length of the lens to its effective aperture (f/D). Because it is

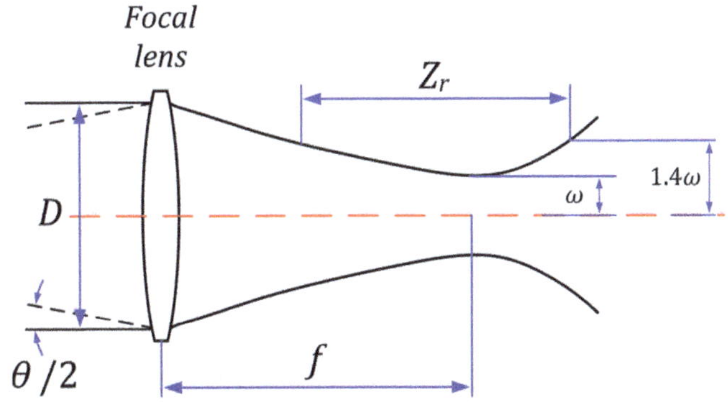

Figure 4.5. Focusing of a Gaussian beam, drawn based on [27].

impractical to work with an F-number less than unity, the minimum value of the spot size will be of some order of magnitude as λ [27].

The distance over which the focused beam has about the same intensity is the depth of focus, and it is defined as the distance over which the focal spot size changes between –5% and 5%. Depth of focus lies between the 1.4ω spot size points or two times the Rayleigh range. In machining, one must provide sufficient depth of focus to compensate for vibration and for any inaccuracy in the positioning of a workpiece [13]. The depth of focus Z_f is approximately:

$$Z_f = 2Z_r = 2\pi\lambda F\#^2 = 2\pi\lambda\left(\frac{f}{D}\right)^2 \tag{4.8}$$

where Z_f is the depth of focus and Z_r is the Rayleigh range.

The Rayleigh range is the distance from the center of that area where the beam is parallel (the waist) to the position where the beam cross section diameter has doubled. Equations (4.6) and (4.7) show that the laser wavelength has an important role in decreasing the spot size and consequently increasing the beam intensity. In addition, it is evident that laser beams of shorter wavelength have more limited depth of focus and Rayleigh range [28, 29].

4.3 Laser beam polarization and angles of incidence

Polarization describes the direction at which the electric field of a laser beam oscillates in relation to the vector of laser beam propagation. The key factor of laser machining of materials is the electric field vector orientation and temporal stability, which is represented by the type of polarization. Laser polarization may be linear where the electric field vector oscillation is confined in a plane along the propagation of the light with fixed value. When the tip of the vector oscillation makes a circle, with the same magnitude, or ellipse, with changing magnitude, the polarization is called circular or elliptical, respectively [30]. The direction of the electric field with respect to the plane of incidence determine the type of linear polarization. The linearly polarized light could be of p-polarization state when the electric field is parallel to the plane of incidence. While, in s-polarization the electric field is perpendicular to the plane of incidence as illustrated in figure 4.6(a) [30]. The importance of polarization in laser–matter interaction is represented by the portions

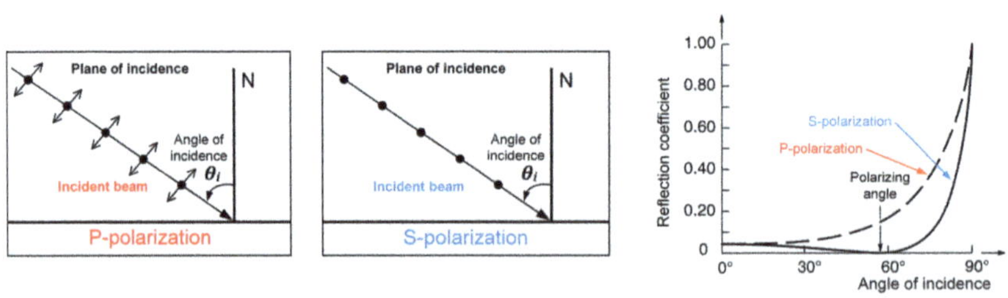

(a) (b)

Figure 4.6. (a) Linear polarization states, drawn based on reading [30] and (b) reflectivity of linear polarized light at air–glass ($n = 1.5$) as a function to the angle of incidence, modified based on [25], John Wiley & Sons, copyright (2009).

of absorbed and reflected light by the processed material [31]. Figure 4.6(b) shows that the incident laser beam on the surface of glass is not necessarily employed completely in the interaction, where part of it will be lost due to reflection, which is highly dependent on the material type, wavelength and angle of incidence. The loss in the s-polarized laser beam due to reflectivity is more significant than the p-polarized beam [25, 32].

Based on Fresnel equations, the absorptivity to a linearly polarized light that is incident on a solid surface as a function of polarization state and angle of incidence can be expressed as:

$$A_p = \frac{4n \cos \theta}{(n^2 + k^2)\cos^2 \theta + 2n \cos \theta + 1} \tag{4.9}$$

$$A_s = \frac{4n \cos \theta}{n^2 + k^2 + \cos^2 \theta + 2n \cos \theta + \cos^2 \theta} \tag{4.10}$$

where A_p and A_s are the absorptivity for p-polarized and s-polarized beams by the material, respectively, and θ is the angle of incidence.

Equations (4.9) and (4.10) are valid when the condition ($n^2 + k^2 \geqslant 1$) is satisfied. This is possible for metals when processed by a laser beam of wavelength 0.5 μm or greater. Losses in the optical power of a laser beam are unavoidable but can be minimized by manipulating the state of polarization or the angle of incidence. For normal incidence ($\theta = 0$) the case becomes polarization-independent and equations (4.9) or (4.10) become like equation (4.4) [33]. Increasing the absorptivity can be achieved when the laser beam is directed on the material surface with, or near, the Brewster angle for different laser material processing applications. In general, the emergent beams from CO_2 and diode lasers are linearly polarized while for Nd:YAG the beam is randomly polarized [31, 34].

Machining materials with a linearly polarized laser beam is a direction-dependent process affecting the quality, resolution and dimensional tolerance. This is noticeable in orthogonal machining of linear paths where the machined groove width increases up to 30% when two perpendicular dimensions are compared [35]. In addition, machining of circular or helical patterns show perpendicular elongation to the direction of polarization at the exit side [34]. Inside the groove, low radiation absorption by the side walls occur when a p-polarized laser beam is used resulting in a V-shaped profile. The drawbacks of machining with linearly polarized beams can be overcome with other types of polarized laser beams such as circular, tangential, radial or random. Circular polarization takes place when the electric field vector rotates uniformly in a circle in a fixed value [36]. Random polarization of a laser beam is attributed to random orientation of the plane of oscillation with time. The absorptivity when a circular polarized laser beam is used can be expressed as the average value of p- and s-polarization states:

$$A_c = \frac{1}{2}(A_p + A_s) \tag{4.11}$$

where A_c is the absorptivity for circular polarized light [33].

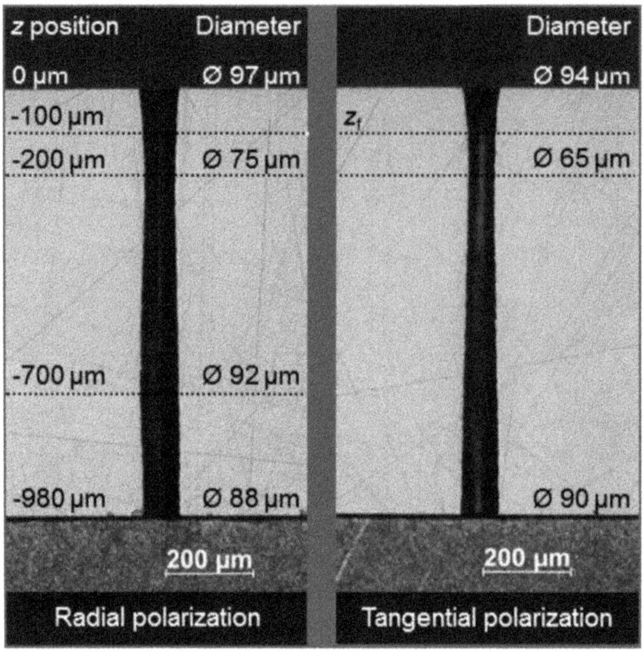

Figure 4.7. Microdrilling with picosecond laser pulses of 1 mm thick CrNi–steel alloy using two polarization states; radial and tangential. Reprinted from [9], copyright (2011), with permission from Elsevier.

Processing the same workpiece with different polarization states produces various features of geometry, dimensions and quality as can be seen in figure 4.7 [9, 36].

4.4 Pulse duration and pulse repetition rate

The effect of pulse duration has been discussed in section 3.2 in terms of its role in determining the interaction regime with respect to characteristic temporal constants of the material (the electron cooling time and lattice heating time). The pulse duration, which refers to the full width at half maximum (FWHM) of the amplitude, is a critical consideration in precision machining processes. Broadly speaking, in contrast to shorter pulses, long pulses allow more thermal interactions, which lead to thermal damage and a wider HAZ. However, as the pulse duration is shortened, the peak power rises, and as a result a higher portion of energy will be above the threshold needed for machining. In micro-/nano-machining applications, shorter pulses mean more laser energy being consumed in vaporizing and ejecting machined material with less molten material. Consequently, less energy is transformed and less heat flows towards the surrounding material [37].

Although the pulse duration τ primarily represents the interaction time, it also reflects the energy contained within the pulse:

$$E_p = P_p\tau \tag{4.12}$$

where E_p is the pulse energy and P_p is the peak power of the pulse.

In addition, the depth of penetrated heat in the substrate has a direct relationship with the pulse duration. In other words, in the absence of any other factors, increasing the pulse duration means more heat to diffuse into the surrounding material, producing larger thermal effects; while shorter pulses reduce thermal damage and offer significant advantages in terms of recast, cracking, HAZ and dross [38, 39].

Steen and Mazumder have classified short pulses into fundamental types according to the manner of material removal. The first class of short pulses of duration less than 1 ns lead to thermal equilibrium and the majority of heat loss by vaporization is achieved. The other class is for ultra-short pulses of less than 1 ps duration that lead to non-thermal equilibrium between the electrons and lattice where heat loss by direct vaporization is dominant [2].

The number of pulses delivered to the target per second is called the pulse repletion rate (PRR). The inverse of PRR is the pulse repetition time (PRT), which represents the time spacing between sequential pulses (PRR = 1/PRT). However, pulsed laser beams with high or extreme PRR show a closer effect to the CW effect on the target. The pulse duration and pulse repetition rate determine the rate of energy pouring on the target, which is termed the average peak power (P_{av}):

$$P_{\text{ave}} = \tau \cdot \text{PRR} \tag{4.13}$$

For a successful ablation process and energy conservation, PRR should be high enough to maintain the temperature required for ablation and avoid target cooling during the process. As these parameters are highly dependent on the optical, thermal properties of a material as well as process conditions, there is no solid rule to determine how the optimum machining process can be conducted [2, 40]. For case studies, the reader can refer to [41–45].

4.5 Laser beam transverse electromagnetic mode (TEM)

TEM describes the pattern of energy distribution over the cross section of a laser beam, as shown in figure 4.8. This parameter is designated by TEM_{mn}, where m and

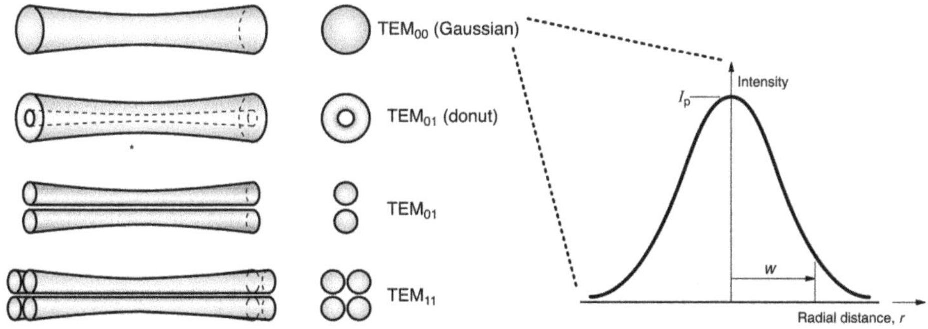

Figure 4.8. TEM_{00} versus higher multi transverse modes, modified based on [25]. John Wiley & Sons, copyright (2009).

n are two notations used to indicate the number of modes, across the cross section, in a plane perpendicular to the direction of the beam propagation [25]. A Gaussian beam is the ideal laser beam mode, which has a radially symmetric cross section of greatest intensity at the center and tails off at the edges. The Gaussian beam distribution is termed the fundamental mode or TEM_{00}. Most laser applications are virtually designed to employ TEM_{00} due its ability to be focused into the smallest spot possible, compared to other transverse modes, and thus the highest intensity can be achieved. One of the precision determinant parameters in laser machining is the use of TEM_{00}; it can be thought of as the sharpest machining tool. In other higher modes, the beam spreads out resulting in larger spot sizes and consequently lower intensities. With TEM_{00}, the possible precise machining contributes to increasing the scanning speed and results in narrower machined width.

4.6 Pulse shape

A single pulse life cycle starts with gradual heating, regular heating and then gradual cooling at the end of the pulse. Figure 4.9(a) depicts a rectangular pulse shape type, which is characterized by one energy sector and a constant peak power over its duration. This type of pulse shape is the simplest and most desired pulse type in research and industrial applications. In modern laser devices, the pulse shape is designed by the user and fed to the flash lamp of the system. Due to the nature of the electrical components, the realistic shape of the pulse is as illustrated in figure 4.9(b). To alter the pulse shape, the user should design the segments that compose the pulse using a process called pulse shaping. The latter is the process of energy temporal redistribution inside a single pulse. The pulse shaping technique is based on dividing a single pulse into many individual sectors for defined values of peak power and duration. The interaction behavior between laser and material can be changed by manipulating the distribution of energy within the pulse time [46]. Case studies of the pulse shape effect are presented in these articles [47–51].

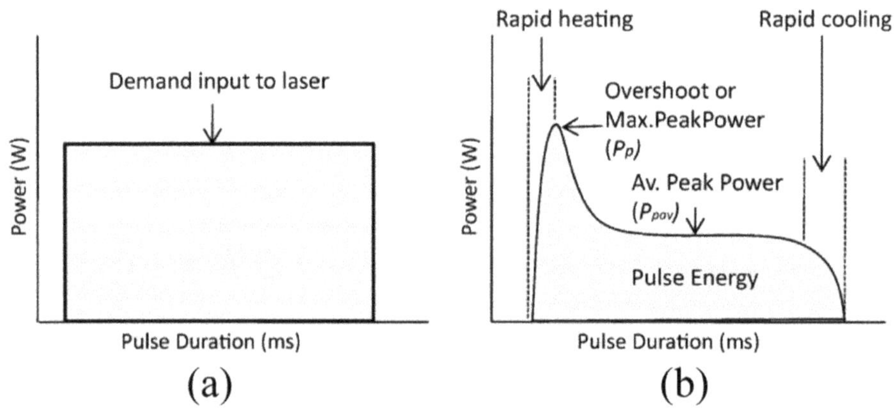

Figure 4.9. A single pulse life cycle: (a) square pulse and (b) real pulse shape, modified based on. Reprinted from [46], copyright (2010), with permission from Elsevier.

4.7 Laser beam intensity and peak power

Laser beam intensity is known as the peak power (P_p) per unit area. For a Gaussian beam, the power distribution is maximum at the center of the beam and decreases exponentially toward the circumference as expressed in the following equation:

$$I(r) = I_o \exp\left(\frac{-2\,r^2}{w^2}\right) \tag{4.14}$$

and

$$I_o = \frac{2\,P_p}{\pi w^2} \tag{4.15}$$

where $I(r)$ is the intensity at a given radius on the spot, I_o is the peak intensity, r is a given radius within the spot, and w is radius at which the beam intensity falls to $1/e^2$ of the peak intensity (0.14 I_o). The term $1/e^2$ of the peak intensity means 86% of the spot energy is within the spot area of radius w.

The temporal distribution of the intensity within a Gaussian laser pulse is given by:

$$I(t) = I_o \exp\left(\frac{-4\ln(2)t^2}{\tau^2}\right) \tag{4.16}$$

where $I(t)$ is the intensity after a given time t and τ is the pulse duration corresponding to the FWHM.

All materials will be directly vaporized with almost no melting when the beam intensity is larger than 10^{10} W cm^{-2} [34]. The required time (t_{rq}) to reach the vaporization phase is correlated to the laser beam intensity. For laser beam intensities of less than 10^8 W cm^{-2}, light absorption by the generated vapor is not significant and can be considered transparent to the incident radiation [52]. Higher intensities mean less time required to reach the vaporization phase as given in the following relation [49]:

$$t_{rq} \approx D\left(\frac{\Omega\,\rho}{(1 - R)I}\right)^2 \tag{4.17}$$

where D is the thermal diffusivity of the material, ρ is the density, Ω is the enthalpy of vaporization per unit of mass and I is the laser beam intensity.

In general, higher values of laser beam intensity, in combination with shorter pulse duration, contribute to more material removal via ablation and minimized thermal damage that is inflicted on the surrounding area to the target. The pulse duration has an inverse proportionality with the laser beam intensity. Therefore, the threshold intensity for ablation has an inverse proportionality with pulse duration [49]:

$$I_{th} = \frac{\Omega\rho\sqrt{D}}{(1 - R)\sqrt{\tau}} \tag{4.18}$$

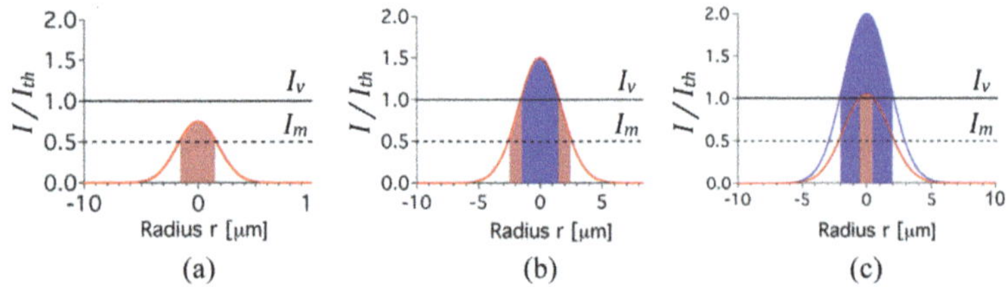

Figure 4.10. Beam intensity effect on the extension of melted and ablation zones: (a) extension of the melt zone, (b) extension of the ablated zone and (c) comparison between the extension of both melted and ablated regions, modified based on. Reprinted from [56], copyright (2015), with permission from Elsevier.

In general, with shorter pulse durations, it is necessary to have a sufficient laser intensity more than the ablation intensity to yield ablations at larger depths. The width of the ablated region can be controlled by manipulating the value of the beam intensity. Figure 4.10 shows this dependence for a Gaussian beam focused to a defined spot area on the target surface. For a known ablation threshold intensity I_{th} of a material, the ratio of the applied intensity to the threshold intensity for ablation (I/I_{th}) is in a relationship with the processed width diameter. Figure 4.10(a) shows the extent of the melt when the ratio is less than one ($I < I_{th}$). When the applied intensity is increased above the threshold intensity for ablation ($I > I_{th}$), the width becomes wider due to the existence of the vapor and melt extent as shown in figure 4.10(b). The latent heat for fusion is less than that for vaporization, therefore the processed region width with melting is wider than with vaporization. It can be deduced that the processed width can be decreased with decreasing the intensity of the applied beam. Figure 4.10(c) shows that to achieve less width, it is better when the applied intensity is close to the threshold intensity for ablation. For more research articles tackling this effect, the reader can refer to [53–55].

4.8 Fluence

Fluence, which is also called energy density, is the amount of energy of a single pulse per area of $1/e^2$ magnitude of the spot size. In most of the literature, fluence is commonly measured in Joule per unit area of square centimeter (J cm^{-2}). Unlike intensity, fluence does not include the temporal behavior of energy input during specific pulse duration. In other words, only the spatial distribution of energy is described by the fluence. For example, a fluence of 16 mJ cm^{-2} is the same value for two different sets of parameters having the same spot size; the first one having 10 ns pulse duration and 10 kW peak power and the second having 10 ps pulse duration and 1 MW peak power [50]. Fluence can be expressed by the following relations [57]:

$$E_p = \int_{-\infty}^{\infty} \int_{-\infty}^{\infty} F(x, y) dx dy \tag{4.19}$$

$$F(x, y) = \int_{-\infty}^{\infty} I(x, y, t) dt \tag{4.20}$$

where E_p is the pulse energy, x and y are the orthogonal coordinates in the plane perpendicular to the radiation direction in the z-axis and t is the time.

Like intensity, the spatial distribution of the fluence is maximum at the center of the beam and decreases exponentially towards the circumference based on the following equation:

$$F(r) = F_o \exp\left(\frac{-2\, r^2}{w^2}\right) \tag{4.21}$$

where $F(r)$ is the fluence at a given radius on the spot and F_o is the peak fluence, r and w as previously defined for intensity calculations in equations (4.14) and (4.15).

The fluence of a Gaussian beam of radius w is expressed as the fluence that falls within $1/e^2$ of the peak fluence (0.14 F_o) and is given by the following equation:

$$F(w) = F_o \exp(-2) \tag{4.22}$$

The peak fluence is obtained from the solution of equation (4.19) by knowing the energy of the pulse and the beam radius (or spot size) [57]:

$$F_o = \frac{2\, E_p}{\pi w^2} \tag{4.23}$$

By employing the intensity distribution in equation (4.16) and the solution approach from equation (4.15), the dependence of the peak intensity with the peak fluence can be obtained from:

$$I_o = \frac{2\, F_o}{\tau}\sqrt{\frac{\ln(2)}{\pi}} \tag{4.24}$$

The peak intensity as a function of pulse energy for a Gaussian or elliptical beam of known area A can be calculated from substituting equation (4.23) in equation (4.24):

$$I_o = \frac{4}{\tau A}\sqrt{\frac{\ln(2)}{\pi}}\, E_p \tag{4.25}$$

The importance of intensity in laser machining of materials was clarified in sections 4.5–4.7 in terms of determining the minimum level (I_{th}) required to achieve ablation for a specific depth. Similarly, when the applied fluence exceeds the threshold fluence (F_{th}) ablation starts. Also, similar to the analysis presented with intensity, the dimensions of the resulting feature, i.e. width and depth, depends on the ratio of F/F_{th} and other process conditions. For a fixed irradiation, the diameter of the impact (D_m) is a function for the laser maximum and threshold fluence and number of incident shots on the target and is given by the following equation [58]:

$$D_m = \sqrt{2\, w^2 \ln\left(\frac{F_o}{F_{th(N)}}\right)} \tag{4.26}$$

where N is the number of laser pulses shots.

When ablating materials with nanosecond pulses, the produced feature has a direct relationship with the level of fluence. From a low level of fluence up to threshold fluence results in a slight impact represented by roughening the surface of the target. When the fluence exceeds the threshold fluence, the relation between the ablation depth and fluence becomes linear. At higher fluence values, the ablated material shields the target leading to less ablation depth, in this case the relation between the depth and fluence is logarithmic [59]. For more case studies on the effect of fluence, the reader can refer to [17, 60, 61].

4.9 Scanning speed

Recently, high-precision laser micro-/nano-machining systems have offered exact control for each pulse on the target within high repetition and continuous train of laser pulses. Tuning of high repetition rates with the laser spot speed on the target can harness the energy input on the target among pulses towards improving the quality of the process [62, 63]. Laser machining with high precision and resolution requires homogenous ablation of the target along with a specified path. This can be achieved by controlling the speed, which is in turn related to the resulting spots overlapping on the target surface, as illustrated in figure 4.11. For a linear scanning speed, the pulse overlap can be estimated by the following equation:

$$O_v = \frac{D - x}{D} \times 100 \tag{4.27}$$

where O_v is the percentage of pulse overlap, x is the overlap distance parameter and D is the spot diameter equal to twice the spot size (w). By considering the scanning speed v and the pulse repetition rate PRR, equation (4.27) can be rewritten in the following form [64, 65]:

$$O_v = \left(1 - \frac{v}{2w \cdot \mathrm{PRR}}\right) \times 100 \tag{4.28}$$

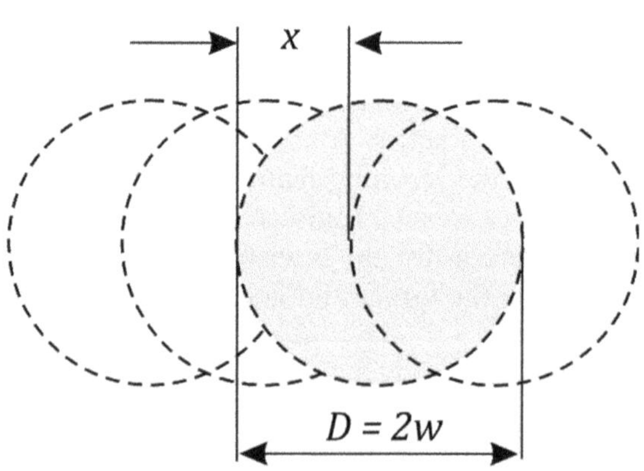

Figure 4.11. Laser pulses overlapping, modified based on. Reprinted from [68], copyright Optica Applicata.

Apparently, increasing the scanning speed, or reducing the pulse repetition rate, leads to a low overlap percentage of pulses. In addition to the efficiency of the ablation process, pulse overlap influences the surface roughness of the ablated target [64]. A higher value of overlap gives more deposition of energy on the spot area. By rearrangement of equation (4.27) with the definition of the duty cycle in equation (4.29), the number of overlapping spots irradiated on the same location on the target (N) can be obtained [66, 67]:

$$\mathrm{DC} = \frac{\tau}{\mathrm{PRR}} \tag{4.29}$$

$$N = \frac{1}{1 - \mathrm{PRR}} = \frac{2w \cdot \mathrm{PRR}}{v} = \frac{2\,w}{v \cdot \mathrm{PRR}} + \mathrm{DC} \tag{4.30}$$

At high pulse repetition rates and even at high scanning speeds, a given area can be irradiated with a sufficient number of pulses to cause material removal via ablation. Pulse overlapping offers multiple laser pulses incidence on the same spot giving the possibility of ablation with employing fluences less than those of a single pulse ablation [69].

The threshold fluence (F_{th}) value required to cause ablation using a single pulse decreases dramatically when irradiating the target surface with N number of pulses having the same pulse energy. The surface damage threshold drops after the first laser shots until reaching an almost constant level. This reduction for the first shots is attributed to the formed defect due to laser-induced damage [70]. The increase in absorption due to a formed defect from the first shots results in a decrease in the ablation threshold, this behavior is called 'incubation'. The effect of the latter on the ablation threshold is represented as follows [69]:

$$F_{th}(N) = F_{th}(1)\, N^{\xi-1} \tag{4.31}$$

where ξ is the incubation factor.

From equations (4.28) and (4.31), the average thermal input per unit area (E_a) delivered to the target for a given laser spot size can be deduced as follows [66]:

$$E_a(N) = E_a(1)\left(\frac{v}{D \cdot \mathrm{PRR}}\right)^{\xi-1} = E_a(1)\left(\frac{1}{1 - O_v}\right)^{\xi-1} \tag{4.32}$$

Some research works that focused on the effect of overlapping and scanning speed are presented in [17, 71–76].

4.10 Assist gas, type and flow rate (or pressure)

In laser micro-/nano-machining of materials, melting or/and vaporization allows for material removal with or without the use of assist gas [77]. Sometimes, laser machining is accompanied by high pressure assist gas through a gas nozzle to enhance the process. Assist gas contributes to the enhancement of laser machining in different ways. Cooling the processed area, blowing out the molten and vaporized material, and reducing the spattered materials on the surface of the processed material are the main benefits. Another important advantage is the protection of the laser system's optical components from the harsh effects of the ejected droplets and contaminates [78, 79]. The two common configurations for gas pumping, through gas nozzles, are the coaxial and off-axis types. In the coaxial type, the gas is directly blown along the same axis of the laser beam direction onto the surface of the target. Many past research works considered different parameters in connection with the assist gas effect on the machining quality. These parameters included the gas type, gas pressure, velocity and nozzle design. Many types of assist gases such as helium, argon, nitrogen, oxygen and air can be employed depending on the reactivity of the gas with the processed material. For example, air as a cheap gas, is a suitable choice as an assist gas in processing most polymers. Nitrogen and inert gases like argon are suitable for blowing debris and vapors, ejection of melt sidewall burning effects, avoiding re-solidification formation at the edges and building a tent over the interaction zone to avoid oxidation. With stainless steel, nitrogen is more suitable than oxygen to avoid the oxidation problems. For the same reason, argon is preferable with titanium processing due to its inertness to interact with the metal melt. Irrespective of the cost, to increase the efficiency, some researchers have used oxygen to apply additional energy to the processed zone through exothermic interaction [25, 80]. Different researchers have studied the effect of assist gas type, pressure and flow rate; these studies can be found in [79, 81–84].

4.11 Focus position

The distance between the focusing lens and workpiece surface is called the focus position. There are three possible locations of minimum spot size 'the focal spot' relative to the workpiece during micromachining. These locations are either on the surface of the workpiece, below or above the upper surface. The focus position is highly correlated with the laser beam intensity on the surface. It has an effective impact on the quality, profile and volume of the ablated material during laser ablation [85]. Sometimes the focus position deviates from the ideal position due to errors in the mechanical or optical systems resulting in degradation of the accuracy quality [86]. Ideally, to maintain the ablation process with the same efficiency, the deviation in focal position should not exceed the depth of focus length [87]. Wang

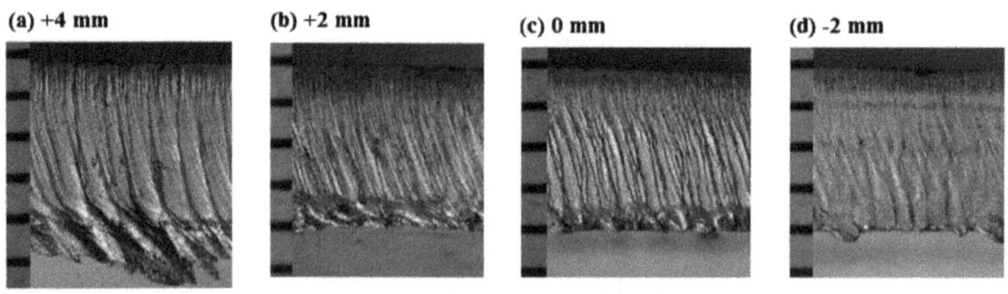

Figure 4.12. Effect of focal point position on striation pattern (aluminum 4 mm: 4 kW, 7 m min^{-1}, and 20 bar nitrogen assist gas pressure), modified based on. Adapted from [89] with permission of Hindawi (CC BY 3.0).

et al showed that the laser focus position represents an important factor for the morphology enhancement and ablation quality. Matching between laser fluence and focus position for producing high quality microstructures can be attained for the same focusing lens [88]. In another work by Wandera *et al* on laser metal-cutting, it was shown that the machining width, quality and morphology differ with control of the focus position. Figure 4.12 shows images of 4-mm thick aluminum workpieces machined with different values of focus position [89].

4.12 Summary

This chapter introduced and discussed the influence of processing and laser beam parameters affecting micro- and nano-laser machining applications. The chapter begins by a general overview of the parameters considered in most research works in the field and that are regarded as most influential. These parameters are categorized into laser beam parameters, material-related parameters and processing parameters. The chapter, thereafter, gives a deeper discussion of definitions, nature, effects, mathematical models and calculations related to each of these parameters. These discussions and illustrations are supported by results demonstrated in several research works in the literature. The knowledge presented in this chapter sheds light on several laser–material interaction regimes with mathematical modeling. This includes presentation of definitions and formulae to calculate breakdown thresholds, pulse energy, beam power, energy deposition, beam polarization, absorptivity, laser intensity and fluence, etc. Furthermore, simplified and critical presentation of workpiece scanning, processing and optical delivery system parameters is included. Thus, the chapter forms an excellent and comprehensive start for researchers who aim to plan and carry out experimental research in the field of micro- and nano-machining. The chapter is also a good quick reference for researchers who are familiar with the laser machining field.

References

[1] Majumdar J D and Manna I 2003 Laser processing of materials *Sadhana* **28** 495–562

[2] Steen W M and Mazumder J 2010 *Laser Material Processing* 4th edn (London: Springer)

[3] Chaudhary K, Rizvi S Z H and Ali J 2016 Laser-induced plasma and its applications *Plasma Science and Technology—Progress in Physical States and Chemical Reactions* ed T Mieno (Rijeka: InTechOpen) pp 259–91

[4] Zhou R, Zhang Z and Hong M 2020 The art of laser ablation in aeroengine: the crown jewel of modern industry *J. Appl. Phys.* **127** 080902

[5] Dumitras D C 2012 *Nd:YAG Laser* ed D C Dumitras (Rijeka: InTech)

[6] Klein R J 2011 *Welding: Processes, Quality, and Applications* (New York: Nova Science)

[7] Cao X, Jahazi M, Immarigeon J and Wallace W 2006 A review of laser welding techniques for magnesium alloys *J. Mater. Process. Technol.* **171** 188–204

[8] Al-Najjar F, Akman E, Genc Oztoprak B, Gunes M, Gundogdu O, Kacar E, Hajim K I and Demir A 2013 Evaluation of PMMA joining to stainless steel 304 using pulsed Nd:YAG laser *Opt. Laser Technol.* **49** 143–52

[9] Weber R, Michalowski A, Abdou-Ahmed M, Onuseit V, Rominger V, Kraus M and Graf T 2011 Effects of radial and tangential polarization in laser material processing *Phys. Proc.* **12** 21–30

[10] Bertrand C and Poulon-Quintin A 2011 Effect of temporal pulse shaping on the reduction of laser weld defects in a Pd–Ag–Sn dental alloy *Dent. Mater.* **27** e43–50

[11] Jensen M F 2004 Laser micromachining of polymers *Ph.D. Thesis* Technical University of Denmark (DTU)

[12] Toyserkani E, Khajepour A and Corbin S 2005 *Laser Cladding* (Boca Raton, FL: CRC Press)

[13] Singh S K and Maurya A K 2017 Review on laser beam machining process parameter optimization, *Int. J. Innov. Res. Sci. Technol.* **3** 34–8

[14] Meijer J, Du K, Gillner A, Hoffmann D, Kovalenko V S, Masuzawa T, Ostendorf A, Poprawe R and Schulz W 2002 Laser machining by short and ultrashort pulses, state of the art and new opportunities in the age of the photons *CIRP Ann.* **51** 531–50

[15] Jaeggi B, Neuenschwander B, Hunziker U, Zuercher J, Meier T, Zimmermann M, Selbmann K H and Hennig G 2012 Ultra high precision surface structuring by synchronizing a galvo scanner with an ultra short pulsed laser system in MOPA arrangement *Laser Applications in Microelectronic and Optoelectronic Manufacturing XVII (San Francisco, CA, 21–26 Jan 2012) Proc. SPIE* vol 8243 pp 109–19

[16] Teixidor D, Thepsonthi T, Ciurana J and Özel T 2012 Nanosecond pulsed laser micromachining of PMMA-based microfluidic channels *J. Manuf. Processes* **14** 435–42

[17] Garasz K, Tanski M, Kocik M, Iordanova E, Yankov G, Karatodorov S and Grozeva M 2016 The effect of process parameters in femtosecond laser micromachining *Bulg. J. Phys.* **43** 110–20

[18] Powell J 1998 *CO_2 Laser Cutting* 2nd edn (Berlin: Springer)

[19] Nakata Y and Ohara K 1994 Laser machining apparatus *European Patent Office Patent EP* 0600097 B1

[20] Biswas S, Ghosh A, Odusanya A and Kietzig A-M 2019 Effect of irradiation wavelength on femtosecond laser-induced homogenous surface structures *Appl. Surf. Sci.* **493** 375–83

[21] Zhao W, Wang W, Li B Q, Jiang G and Mei X 2016 Wavelength effect on hole shapes and morphology evolution during ablation by picosecond laser pulses *Opt. Laser Technol.* **84** 79–86

[22] Oosterbeek R N, Ashforth S, Bodley O and Simpson M C 2017 Ablation threshold dependence on incident wavelength during ultrashort pulsed laser ablation, *Int. J. Nanotechnol.* **14** 313–22

[23] Shan J 2010 Laser welding of magnesium alloys *Welding and Joining of Magnesium Alloys* ed L Liu (Cambridge: Woodhead Publishing Ltd)

[24] Bigot S, Lacan F, Hirshy H, Petkov P, Babenko M, Castro G, Sweeney J, Ugail H and Whiteside B 2014 Micro and nano structuring of sapphire for micro injection process investigation *Proc. 9th Int. Conf. on MicroManufacturing (Nanyang Technological University (NTU) and National University of Singapore (Singapore, 25–28 Mar 2014)*

[25] Kannatey-Asibu J E 2009 *Principles of Laser Materials Processing* (Hoboken, NJ: Wiley)

[26] Zakariyah S S 2012 Laser ablation for polymer waveguide fabrication *Micromachining Techniques for Fabrication of Micro and Nano* ed M Kahrizi (Rijeka: InTechOpen) pp 109–30

[27] Migliore L 1996 *Laser Materials Processing* (New York: Marcel Dekker)

[28] Nath A K 2014 Laser drilling of metallic and nonmetallic substrates *Comprehensive Materials Processing* (Reference Module in Materials Science and Materials Engineering) vol 9 (Amsterdam: Elsevier) pp 115–75

[29] Beckmann L H J F and Ehrlichmann D 1995 Optical systems for high-power laser applications: principles and design aspects *Opt. Quantum Electron.* **27** 1407–25

[30] McClelland A and Mankin M 2018 Introduction *Optical Measurements for Scientists and Engineers* (Cambridge: Cambridge University Press) pp 1–24

[31] Rodrigues G C and Duflou J R 2016 Effects of different polarization strategies on laser cutting with direct diode lasers *Phys. Proc.* **83** 302–9

[32] Eltawahni H A, Benyounis K Y and Olabi A G 2016 High power CO_2 laser cutting for advanced materials—review *Reference Module in Materials Science and Materials Engineering* (New York: Elsevier) pp 1–27

[33] Bergström D 2008 The absorption of laser light by rough metal surfaces *PhD Thesis* Luleå University of Technology, Sweden

[34] Poprawe R 2011 *Tailored Light 2: Laser Application Technology* (Berlin Heidelberg: Springer)

[35] Jagdheesh R, Valarmathi A and Sivakumar M 2015 Influence of laser beam polarization on micromachining of crystalline silicon using ultrafast laser pulses *Lasers Eng.* **32** 19–35

[36] Niziev V G and Nesterov A V 1999 Influence of beam polarization on laser cutting efficiency *J. Phys.* D **32** 1455–61

[37] Schwarz S, Rung S, Esen C and Hellmann R 2018 Influence of pulse duration on high-precision manufacturing of 3D geometries *J. Laser Micro/Nanoeng.* **13** 292–5

[38] Neimash V B, Goushcha A O, Fedorenko L L, Shepelyavyi P Y, Strelchuk V V, Nikolenko A S, Isaiev M V and Kuzmich A G Role of laser power, wavelength, and pulse duration in laser assisted tin-induced crystallization of amorphous silicon *J. Nanomater.* **2018** 1243685

[39] Simonds B J, Meadows H J, Misra S, Ferekides C, Dale P J and Scarpulla M A 2015 Laser processing for thin film chalcogenide photovoltaics: a review and prospectus *J. Photonics Energy* **5** 050999

[40] Brygo F, Dutouquet C, Guern F L, Oltra R, Semerok A and Weulersse J 2006 Laser fluence, repetition rate and pulse duration effects on paint ablation *Appl. Surf. Sci.* **252** 2131–8

[41] Nordin I H W, Okamoto Y, Okada A, Jiang H and Sakagawa T 2016 Effect of wavelength and pulse duration on laser micro-welding of monocrystalline silicon and glass *Appl. Phys.* A **122** 400

[42] Beltrán L, Abbasi H, Rauter G, Friederich N, Cattin P and Zam A 2017 Effect of laser pulse duration on ablation efficiency of hard bone in microseconds regime *Proc. 3rd Int. Conf. on Applications of Optics and Photonics (Faro, Portugal, 8–12 May 2017) Proc. SPIE vol 10453* pp 425–30

[43] Zhang J J, Yang D H, Song C W, Yan Y D and Sun T 2018 Effect of pulse duration on tribological behavior of textured stainless steel by laser surface texturing *Proc. of Asia Int. Conf. on Tribology (Kuching, Sarawak, Malaysia, 17–20)* pp 8–10

[44] Zhang J, Yang D, Rosenkranz A, Zhang J, Zhao L, Song C, Yan Y and Sun T 2019 Laser surface texturing of stainless steel—effect of pulse duration on texture's morphology and frictional response *Adv. Eng. Mater.* **21** 1801016

[45] Auda L H and Al-Hamaoy A R 2020 Effect of multi pulses on glass drilling using CO_2 laser *Technol. Rep. Kansai Univ.* **62** 245–52

[46] Mumtaz K and Hopkinson N 2010 Selective laser melting of thin wall parts using pulse shaping *J. Mater. Process. Technol.* **210** 279–87

[47] Kim H and Lee C 1999 Effect of Nd–YAG laser pulse shape on welding characteristics of STS 310s stainless steel *Sci. Technol. Weld. Joining* **4** 51–7

[48] Kim B-C, Kim T-H, Kim K-B, Kim J-S and Lee H-Y 2002 Investigation on the effect of laser pulse shape during Nd:YAG laser microwelding of thin Al sheet by numerical simulation *Metall. Mater. Trans.* A **33** 1449–59

[49] Deladurantaye P, Cournoyer A, Drolet M, Desbiens L, Lemieux D, BriandMand and Taillon Y 2011 Material micromachining using bursts of high repetition rate picosecond pulses from a fiber laser source *Fiber Lasers VIII: Technology, Systems, and Applications (San Francisco, 22–7 Jan 2011) Proc. SPIE* vol 7914 pp 47–56

[50] Eiselen S, Wu D, Galarneau P and Schmidt M 2013 The role of temporal energy input in laser micro machining using nanosecond pulses *Phys. Proc.* **41** 683–8

[51] Klocke F, Schulz M and Gräfe S 2017 Optimization of the laser hardening process by adapting the intensity distribution to generate a top-hat temperature distribution using freeform optics *Coatings* **7** 77

[52] Grigoropoulos C P 2009 *Transport in Laser Microfabrication: Fundamentals and Applications* (Cambridge: Cambridge University Press)

[53] Ng G K L and Li L 2001 The effect of laser peak power and pulse width on the hole geometry repeatability in laser percussion drilling *Opt. Laser Technol.* **33** 393–402

[54] Li Y, Hu S and Shen J 2014 The effect of peak power and pulse duration for dissimilar welding of brass to stainless steel *Mater. Manuf. Processes* **29** 922–7

[55] Gursel A 2015 Effect of pulse repetition and peak power of Nd:YAG laser for surface treatment on Ti–6Al–4V alloy *Düzce Univ. J. Sci. Technol.* **3** 210–8

[56] Gillner A and Gretzki P 2015 Laser micro-structuring *Micromanufacturing Engineering and Technology* ed Y Qin (Boston, MA: William Andrew Publishing)

[57] Martin S 2005 Damage mechanisms in optical materials by excitation with ultra-short laser pulses *Ph.D. Thesis* Freie Universität Berlin

[58] Žemaitis A, Gaidys M, Brikas M, Gečys P, Račiukaitis G and Gedvilas M 2018 Advanced laser scanning for highly efficient ablation and ultrafast surface structuring: experiment and model *Sci. Rep.* **8** 17376

[59] Black S E 2011 *Laser Ablation: Effects and Applications* (New York: Nova Science)

[60] Singh S S, Khare A and Josh S N 2020 Fabrication of microchannel on polycarbonate below the laser ablation threshold by repeated scan via the second harmonic of Q-switched Nd: YAG laser *J. Manuf. Processes* **55** 359–72

[61] Krishnan R R 2018 Effect of laser fluence on the structural and optical properties of tantalum oxide films ablated by pulsed laser *J. Russ. Laser Res.* **39** 156–64

[62] Rezaei S, Li J and Herman P R 2015 Burst train generator of high energy femtosecond laser pulses for driving heat accumulation effect during micromachining *Opt. Lett.* **40** 2064–7

[63] Nikolov A S, Balchev I I, Nedyalkov N N, Kostadinov I K, Karashanova D B and Atanasova G B 2017 Influence of the laser pulse repetition rate and scanning speed on the morphology of Ag nanostructures fabricated by pulsed laser ablation of solid target in water *Appl. Phys.* A **123** 719

[64] He T, Wei C, Jiang Z, Yu Z, Cao Z and Shao J 2018 Numerical model and experimental demonstration of high precision ablation of pulse CO_2 laser *Chin. Optics Lett.* **16** 041401

[65] Bartolo P, Vasco J, Silva B and Galo C 2006 Laser micromachining for mould manufacturing: I. The influence of operating *Assem. Autom.* **26** 227–34

[66] Jang P-R, Kim C-G, Han G-P, Ko M-C, Kim U-C and Kim H-S 2019 Influence of laser spot scanning speed on micro-polishing of metallic surface using UV nanosecond pulse laser *Int. J. Adv. Manuf. Technol.* **103** 423–31

[67] Tzeng Y-F 1999 Pulsed Nd:YAG laser seam welding of zinc-coated steel *Welding J.* **11** 238–44

[68] Marczak J, Kusiński J, Major R, Rycyk A, Sarzyński A, Strzelec M and Czyż K 2014 Laser interference patterning of diamond-like carbon layers for directed migration and growth of smooth muscle cell depositions *Opt. Appl.* **44** 575–86

[69] Choi H W, Farson D F, Bovatsek J, Arai A and Ashkenasi D 2007 Direct-write patterning of indium–tin–oxide film by high pulse repetition frequency femtosecond laser ablation *Appl. Opt.* **46** 5792–9

[70] Ashkenasi D, Lorenz M, Stoian R and Rosenfeld A 1999 Surface damage threshold and structuring of dielectrics using femtosecond laser pulses: the role of incubation *Appl. Surf. Sci.* **150** 101–6

[71] Al-Najjar F I H 2018 Numerical analysis of the effect of scanning speed on the temperature field distribution for laser heat treatment applications *Al-Nahrain J. Eng. Sci.* **21** 213–22

[72] Benton M, Hossan M R, Konari P R and Gamagedara S 2019 Effect of process parameters and material properties on laser micromachining of microchannels *Micromachines* **10** 123

[73] Wlodarczyk K L, Lopes A A, Blair P, Maroto-Valer M M and Hand D P 2019 Interlaced laser beam scanning: a method enabling an increase in the throughput of ultrafast laser machining of borosilicate glass *J. Manuf. Mater. Process* **3** 14

[74] Erinoshoa M F, Akinlabi E T and Johnson O T 2019 Effect of scanning speed on the surface roughness of laser metal deposited copper on titanium alloy *Mater. Res.* **22** e20190297

[75] Zhao Y, Wang L and He W 2019 Effect of scanning speed on the interface behavior and dendrite growth of laser remelted Fe-based Ni/WC coatings *Coatings* **9** 677

[76] Phala M, Popoola A, Tlotleng M and Pityana S 2018 Effect of laser scanning speed on surface properties of Ti–Si laser clad intermetallic coatings fabricated on Ti–6Al–4V alloy *Int. J. Microstruct. Mater. Prop.* **13** 331–43

[77] Hendricks F, Patel R and Matylitsky V V 2015 Micromachining of bio-absorbable stents with ultra-short pulse lasers *Frontiers in Ultrafast Optics: Biomedical, Scientific, and Industrial Applications XV (San Francisco, CA, 8–10 Feb 2015) Proc. SPIE* vol 9355 pp 1–9

[78] Ho C-C, Shen K-Y, Chen C-S, Chang Y-J, Hsu J-C and Kuo C-L 2017 Swirling gas jet-assisted laser trepanning for a galvanometer-scanned CO_2 laser *Appl. Sci.* **7** 502

[79] Riveiro A, Quintero F, Boutinguiza M, Val J D, Comesaña R, Lusquiños F and Pou J 2019 Laser cutting: a review on the influence of assist gas *Materials* **12** 157

[80] Powell J and Kaplan A 2004 Laser cutting: from first principles to the state of the art *Proc.1st Pacific Int. Conf. on Application of Lasers and Optics (Melbourne, 19–21 Apr 2004)* pp (LMP-LC) 1–6

[81] Grajcar A, Rózañski M, Kamiñska M and Grzegorczyk B 2016 Effect of gas atmosphere on the non-metallic inclusions in laser-welded trip steel with Al and Si additions *Mater. Technol.* **50** 945–50

[82] Mahamood R M and Akinlabi E T 2017 Gas flow rate and scanning speed influence on microstructure and microhardness property of laser metal deposited titanium-alloy *ICMMSE 2017: 2nd Int. Conf. on Mechanics, Materials and Structural Engineering* (Advances in Engineering Research (AER) vol 102) pp 102–8

[83] Sicong L, Bo C, Di B, Shenghong Y and Hong Z 2019 Effect of N_2 shielding gas flow rate on microstructure and weld surface corrosion resistance of high nitrogen steel by laser-arc hybrid welding *Mater. Res. Express* **6** 0865j2

[84] Sun J, Zhao Y, Yang L, Zhao X, Qu W and Yu T 2019 Effect of shielding gas flow rate on cladding quality of direct laser fabrication AISI 316L stainless steel *J. Manuf. Processes* **48** 51–65

[85] Bordatchev E V and Nikumb S K 2006 Effect of focus position on informational properties of acoustic emission generated by laser–material interactions *Appl. Surf. Sci.* **253** 1122–29

[86] Chen J-X, Lin S-W, Zhou X-L and Tuc Y-L 2017 An on-machine error calibration method for a laser micromachining tool *Precis. Eng.* **47** 239–48

[87] Wei A-C, Sze J-R and Chern J-L 2010 Designs for optimizing depth of focus and spot size for UV laser ablation *Appl. Phys.* A **101** 411–6

[88] Wang W, Mei X and Jiang G 2009 Control of microstructure shape and morphology in femtosecond laser ablation of imprint rollers *Int. J. Adv. Manuf. Technol.* **41** 504–12

[89] Wandera C, Salminen A and Kujanpaa V 2009 Inert gas cutting of thick-section stainless steel and medium-section aluminum using a high power fiber laser *J. Laser Applic.* **21** 154–61

IOP Publishing

Laser Micro- and Nano-Scale Processing

Fundamentals and applications

Ahmed Issa and Dermot Brabazon

Chapter 5

Laser-induced modification of surface properties by micro- and nano-scale processing

Chiara Mandolfino, Enrico Lertora, Marco Pizzorni and Carla Gambaro

5.1 Introduction

Laser processing presents unique advantages, which particularly fit the purpose of surface modification. First of all, there is the possibility of providing high energy and at very concentrated levels, with high processing speed, thus creating a short interaction time. Due to this property it is possible to use it to manufacture a wide range of materials and processes. Moreover, the laser fits perfectly into current environmental policies increasingly pushed to replace traditional technologies with 'green' ones, for its environmentally friendly and clean processing capabilities [1, 2]. The typical applications of lasers on bulk material are purely welding, cutting, deposition and forming [3]. The interaction process between a laser and surface layer has been exploited for various uses, such as the modification of microstructure cladding and surface texturing. Successful applications of laser surface processing are: improvement of wear, corrosion and oxidation resistance of many metallic alloys, such as magnesium, titanium, steel and aluminum [4–7].

Materials of various kinds, from metals to polymers, can be selected in a wide range of mechanical properties of the bulk, which make them suitable for the most varied uses. Very often, the characteristics of the bulk material do not correspond to optimal surface characteristics. The surface of a piece must be studied and worked with particular attention, mainly for two reasons. The first is that surfaces are subjected to greater stresses and more direct environmental impact than the inner parts; hence, when such destructive stresses overcome the material resistance limit, fracture, fatigue, wear and corrosion failures may occur on the surface. Consequently, material surfaces are generally modified to make them resistant to the specific environment or stress in which they are used to achieve maximum benefits. The method of treating materials to enhance their surface functionality and make them unaffected by their surroundings is called surface modification or surface

doi:10.1088/978-0-7503-1683-5ch5

engineering. In the literature, several studies on this purpose are reported and they are summarized in [8–11].

The second key factor is that the surface is the contact point with the outside, through which quality joints can be made with other parts. In this regard, in recent years, adhesive bonding has been identified as a technology increasingly capable of replacing more consolidated assembly techniques, such as soldering/brazing, welding and mechanical fastening, and ensuring excellent levels of reliability and resistance over time. Usually, when parts are subjected to severe environmental conditions in service, a bonded assembly is designed to retain a substantial mechanical resistance. For this purpose, pre-treatment of any kind of material is often needed to create sound joints. In order to improve adhesion properties, laser treatment could be an efficient and clean opportunity.

The aim of this chapter is to introduce and go in depth on surface nanostructuring of metallic, polymeric and composite materials by using different laser sources. The superficial changes to focus on are wettability properties and modification of surface roughness. A further analysis is given on how these changes increase the adhesion properties of surfaces for applications in the field of structural bonding. Specific sections of the chapter discuss changes in the durability behavior of adhesive-bonded joints.

5.2 Laser processing of metallic materials for improving surface functionalities

Many studies report the outstanding properties of laser tailored metallic surfaces. In particular, the best results are achieved for materials that require the best performance, such as stainless steels, aluminum and titanium alloys. Also, excellent results are achieved with materials where optimal characteristics of resistance to corrosion are required and durability is an important feature.

5.2.1 Influence of laser modification on surface energy and wettability characteristics

The use of laser processing for surface wettability tuning of metallic materials is an increasing research field [12–18]. Initial work has shown how efficient this recent technique is to achieve highly hydrophobic surfaces (also called 'superhydrophobic surfaces') or, vice versa, to increase surface energy and wettability.

The laser replaces and improves traditional surface processing techniques based on machining. The effect of mechanical processing is maintained in the case of laser processing, but it is also possible to reach the micro- and nano-metric scale. The basic technique of surface wettability tuning consists in changing on one side, the surface chemistry, and on the other side, surface roughness. In particular, surface morphology is one of the key factors determining the wettability of a solid surface. Based on Cassie–Baxter's [19] or Wenzel's [20] theories, the roughness of the sample surface is critical in increasing its hydrophobicity or hydrophilicity. Considering the Wenzel state, with enhanced roughness of a hydrophobic surface, the contact angle (CA) will improve, whereas the CA will reduce as the roughness of a hydrophilic surface increases.

The following relationship describes this set up:

$$\cos \vartheta_a = r \cos \vartheta$$

where ϑ_a is the apparent CA for a rough surface, ϑ represents the CA for a smooth surface and r is the ratio of the real surface area to its flat projected area, which define theoretically the roughness of the material. This state is encountered only where a drop of liquid is in total contact with a surface (figure 5.1a).

This case is quite unrealistic, as usually small pockets of air may remain trapped between the water drop and a hydrophobic rough surface, thus contact is incomplete.

In the 'real' surface case, the drop is in contact on the top of the peaks created by surface roughness and the air is trapped between the peaks and valleys. In this case the relation is given by:

$$\cos \vartheta_a = s_1(\cos \vartheta + 1) - 1$$

where s_1 is the area fraction of the solid surface in contact with the liquid droplet. This CA value increases when the solid fraction decreases.

Although morphology is considered the most important aspect, the mechanism for change of the surface wettability has been found to be dependent on the specific material to be treated, in terms of chemical composition of the surface. Several studies stated that some metallic materials (such as aluminum alloys) report a change in wettability properties after laser treatment also due to the accumulation of different chemical species on the surface. For example, on stainless steel, a certain amount of carbon and its compounds on the laser-treated surface could be found [15]. Similar results were found by Bizi-Bandoki on an experimental aluminum alloy of the 7xxx series. In fact, the amount of carbon on the aluminum surface increased from 55% before treatment to 67% after laser treatment. In particular, two functional groups, the methyl group $-CH_3$ and graphitic carbon appear on the surface and they are found to be responsible for the hydrophobic behavior obtained on the aluminum surface as their presence is only observed on the hydrophobic sample [17].

Based on these theories, several studies have been carried out, reporting the effect of a laser process on the change in surface wettability. In particular, femtosecond

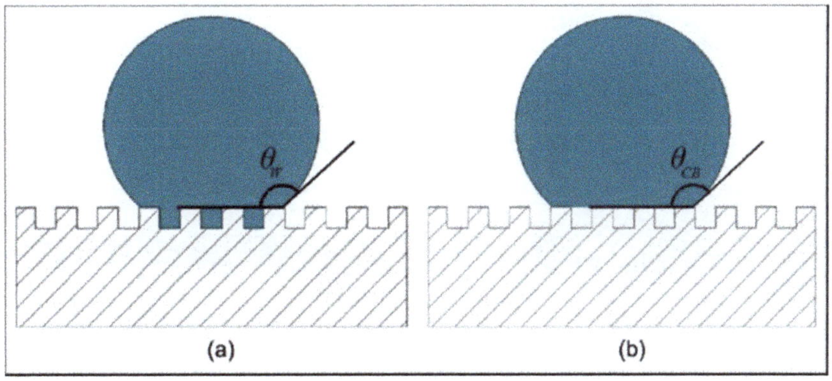

Figure 5.1. (a) Wenzel model and (b) Cassie–Baxter model. (Reproduced from [21], copyright SAGE.)

lasers are used as a roughening method, because of their many advantages. First of all, the laser system can be easily implemented in automated and robotic systems and does not require specific environments, such as vacuum. Then, it could be used on a great range of materials, especially metals, such as stainless steel, titanium or aluminum alloys. In particular, different kinds of microstructures can be obtained, both in micro-/nano-scale levels, such as spikes, holes, bumps or ripples, and surface structures with a periodic trend (figure 5.2).

For example, Bizi-Bandoki *et al* investigated the modification of the topography and the wettability of AISI 316L stainless steel and Ti6Al4V sample surfaces using femtosecond laser texturing [12]. On both materials, the formation of ripples was detected using scanning electron microscopy (SEM) and atomic force microscopy (AFM) in contact mode. Formation of these structures was induced by laser scanning during the ablation regime, along the scanning direction, and were strongly dependent on the number of incident pulses. Both metallic surfaces present originally a hydrophilic behavior. Due to femtosecond laser radiation, their surfaces became hydrophobic, with an increase of about 50° with respect to the nontreated surface.

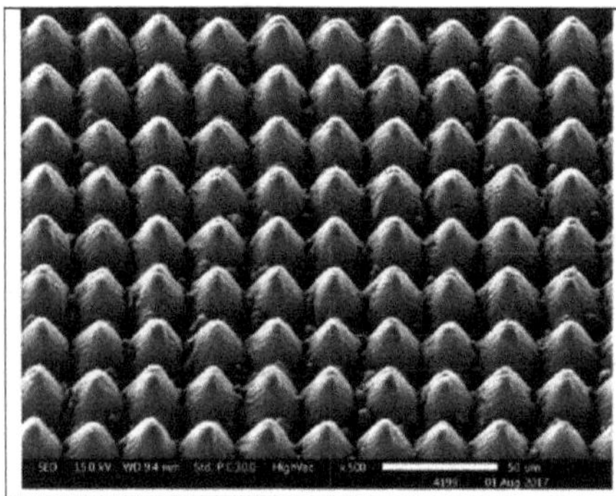

Laser-produced two-scale micro-/nano-bumps array surface texturing. SEM image under 500X magnification. Reproduced by [16]

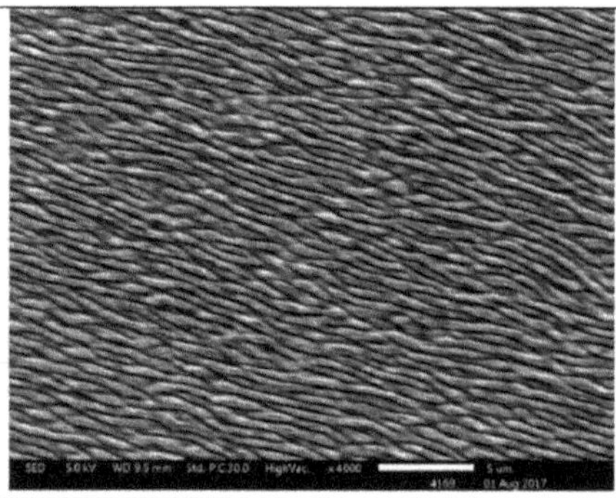

Laser-produced periodic ripple surface textures on a Stavax surface. SEM image under magnification of 1000X; Reproduced by [16]

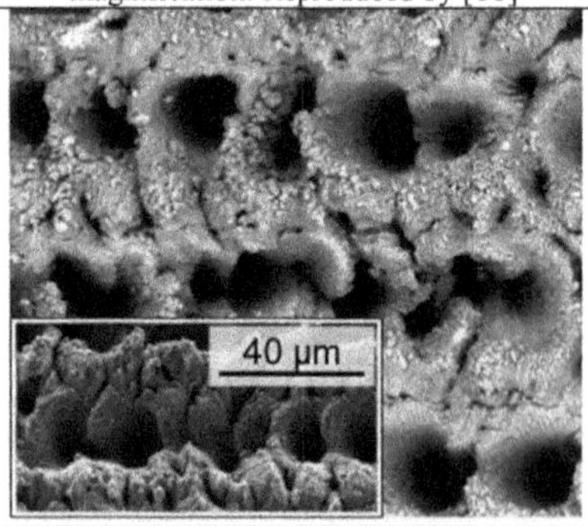

Laser-induced microstructures on titanium. Reproduced by [22]

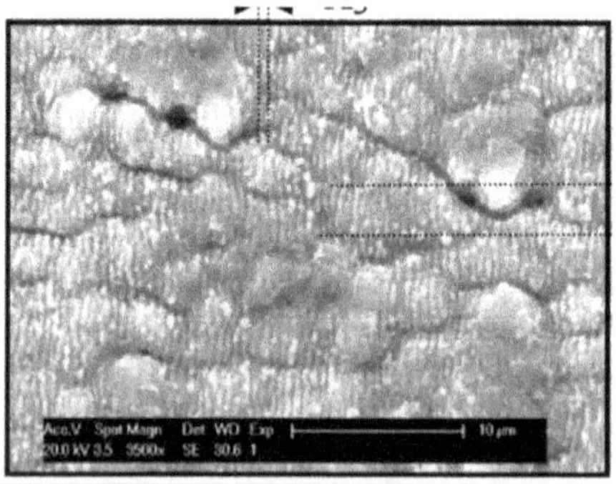

SEM topographic micrographs of Ti6Al4V surfaces. Reproduced by [23]

Figure 5.2. Surface structures obtainable using laser processing. Reprinted from [16], copyright (2018), [22], copyright (2015), [23], copyright (2011), with permission from Elsevier.

Similar results were obtained by Wang *et al* on Stavax Steel, using a picosecond laser surface texturing process. In this case, not only were ripple structures created on the surfaces, but also a two-scale hierarchical 2D array of micro-bumps, and a micro-pits array with nano-ripples [16].

The possibility of optimizing surface texturing of metals as a function of the wettability properties in a wide range of contact angles allows the manufacturing of surfaces suitable for a wide variety of purposes. In the next paragraph some examples will be shown.

5.2.2 Hydrophilicity and hydrophobicity of laser-treated surfaces

The definition of a hydrophilic or hydrophobic surface is determined based on the contact angle of a water droplet on the surface. When the contact angle values are in the range 0° to 90°, the surface is defined as hydrophilic, while for values greater than 90° the surface is defined as hydrophobic. Recent studies have led to the definition of a superhydrophobic surface, when the contact angle is greater than 150° [24–26].

Traditional techniques have been extensively used to produce large-area hydrophilic or hydrophobic surfaces, mainly fabricating micro/nano-structures on the materials, including chemical etching, chemical vapor deposition (CVD), electrochemical deposition and the sol–gel method [27, 28]. In recent years, a strong effort has been made to replace the chemical approaches with the use of chemical-free laser treatment. Being a 'green' technology is not the only advantage. The laser technique offers several technological advantages over the others previously mentioned. It is a highly flexible technique in more than one sense. In fact, it is possible with the same system to obtain different textures by correctly setting parameters and selecting process gases. Moreover, the same system can be implemented for use by micrometer-scale samples up to large surfaces. It is a non-contact process, with the possibility to reduce or eliminate the heat affected zone by tuning the laser pulses and the speed. Last, but not least, the surface structure is usually stable after the process.

Application of hydrophilic metallic surfaces can be found in medical devices, food processing, environmental purification, adhesive bonding, painting, etc [26].

Zhao *et al* created micro- and nano-structures on pure aluminum plates, in order to produce a superhydrophilic surface, using nanosecond laser ablation. By properly adjusting the laser processing parameters, different structures could be obtained, resulting in different hydrophilic effects [29].

The surface of titanium alloy is often modified for medical purposes, and a stable hydrophilic surface is often required, as studied by Fleming *et al* [30].

On the other hand, hydrophobicity is exploited for self-cleaning properties, especially to increase the performance of solar cells, to control bio-adhesion and bio-fouling, and in the food industry [21, 24].

Mainly, femtosecond lasers are used to create the so-called 'lotus-leaf'-like surfaces, replicating the structure of a particularly hydrophobic surface from nature. One of the most interesting recent findings is that, as reported by Kietzig *et al* [18] and Sciancalepore *et al* [15], the superhydrophobic properties of the samples are time-dependent (figure 5.3), at least for stainless steel and titanium alloys. In fact, even if a superhydrophilic behavior is detected immediately after treatment, after

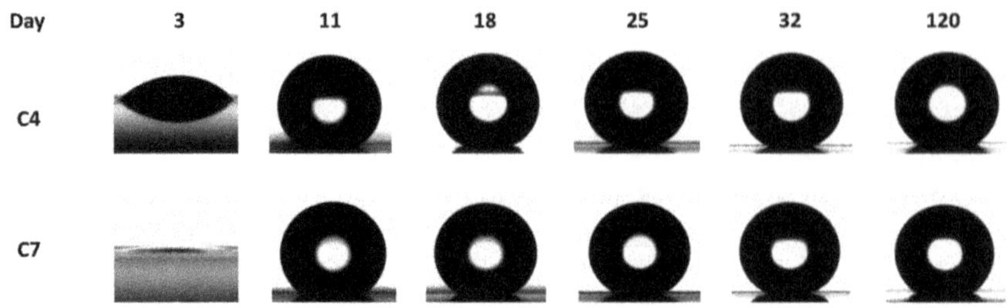

Figure 5.3. Water CA behavior as a function of time for samples treated at a repetition rate of laser pulses of 500 kHz and different energy doses (C4 = 423 J cm^{-2}; C^{-7} = 831 J cm^{-2}). Reproduced from Reprinted from [15], copyright (2018), with permission from Elsevier.

several weeks the contact angle values increased, reaching levels typical of hydrophobic structures.

Kietzig *et al* attribute the time dependency of the wetting behavior changes to the combined effect of surface morphology and surface chemistry, while for Sciancalepore *et al* the results of the evolution of wettability behavior as a function of time demonstrate the effect of the surface chemistry: the laser–material interaction activates the surface in a way that its oxygen content after the laser treatment is high, leading to a hydrophilic state that, however, evolves to hydrophobic, regardless of the specific surface morphology.

5.2.3 Laser texturing to increase adhesive bonding

The ability of the laser process to modify surface morphology and wettability of metal substrates using a clean and effective method has also been exploited to increase the characteristics of structural adhesive-bonded joints. Indeed, adhesion between the adhesive and substrates occurs due to adhesive penetration in surface roughness, in-depth cleaning and the modification of surface chemistry.

The largest number of studies relates to the modification of aluminum and titanium alloys. Wu *et al* conducted an experimental campaign to understand the effect of laser ablation on the performance of adhesive-bonded AA6022-T4 joints. The results show an increase in joint strength by 25%, due to an increase in surface roughness and the formation of a more uniform and thicker aluminum oxide [31].

These findings are substantially confirmed by Zhu *et al* [32], who expanded the study to include cast and extruded parts and by Romoli *et al* [33], who applied critical tensile stress test conditions to the joints.

A further advantage was reported by the investigations mentioned above and the one conducted on aluminum alloy 6061-T6 by Mandolfino *et al* [34], i.e. laser ablation results in a change from adhesive to desirable cohesive failure of the joints (figure 5.4)

The effectiveness of laser ablation treatment on the adhesive bond strength of titanium alloys was also studied, with comparison to traditional chemical surface treatments. Rotella *et al* studied the effect of laser nano-patterning, performed by scanning the laser beam over the entire sample's surface. Results show that laser

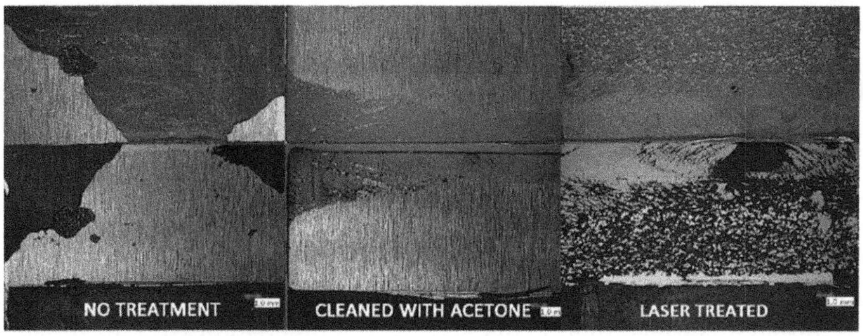

Figure 5.4. Optical images of 6061-T6 alloy fracture surfaces after lap-shear testing. Reprinted from [34], copyright (2015), with permission from Elsevier.

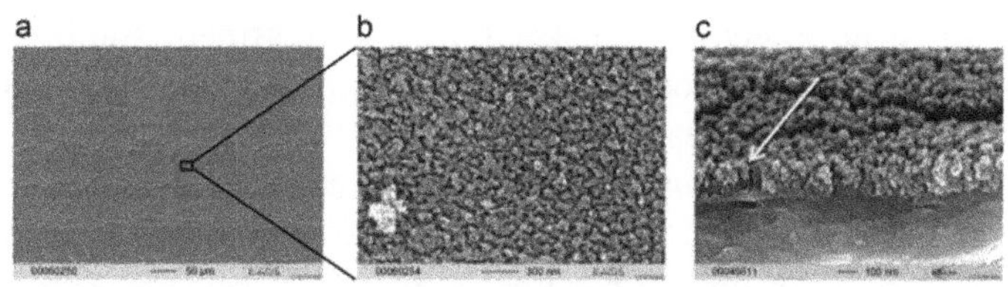

Figure 5.5. SEM micrographs showing (a) top view of the laser pre-treated Ti–6Al–4V surface, (b) high magnification detail and (c) side view cryo fracture of the nano-structured Ti–6Al–4V surface. Reprinted from [36], copyright (2013), with permission from Elsevier.

treatment increased the surface roughness and, consequently, improved joint strength, even after aging in boiling water [35].

Kurtovic *et al* [36] demonstrates that a nano-structured oxide layer could be established exposing Ti6Al4V substrate material to laser irradiation. Good adhesion results were obtained in the wedge tests and the floating roller peeling tests. Figure 5.5 depicts the structuring of the surface on the nanometer-scale obtained with a short-pulse laser system.

Rotella *et al* report also the effect of pulsed laser irradiation on the strength of adhesive joints with different kinds of steel substrates, i.e. dual phase DP500 and stainless steel AISI304, using a pulsed ytterbium doped fiber laser system. In this case the control samples were pre-treated with traditional surface treatment, degreasing and sand blasting. Once the most suitable set of laser processing parameters were selected, single lap-shear and T-peel tests demonstrated that DP steel could withstand intense laser irradiation without major modifications of the surface morphology. It was stated that once a specific level of pulse fluence was achieved, material melting and re-solidification generated micro roughness and increased the real contact area exploitable for bonding. As a result, laser treatment effectively improved the static strength of the joints.

5.2.4 Changes in durability behavior of adhesive-bonded joints

Many researchers agree on the importance of surface pre-treatment as a key factor affecting the durability of adhesive-bonded metallic materials [37, 38]. The majority of studies conducted on metallic substrates report the effect of traditional surface pre-treatments, mainly mechanical and chemical ones. Chemical treatments are considered more effective compared to mechanical ones because they affect both morphology and chemistry of substrates [39].

Recently, as the number of studies on laser treatments to increase the strength of adhesive-bonded joints grows, the effects of this treatment on joint durability is the subject of in-depth studies, mainly on aluminum and titanium substrates [31, 32, 35, 36, 39, 40].

The test methods to assess durability are summarized in [41]. The most used to compare different surface treatments are: water immersion [31, 35], salt spray [40] and the wedge test [36].

The results are generally very good for laser treatments. In particular, in the study of Wu *et al*, an aging of up to six weeks in water soak of laser-treated aluminum substrates exhibited no effect on the strength. The treatment stability is confirmed also by fractography observation; in fact, all the joints made with aged aluminum substrates had a complete cohesive failure.

Even in a salt spray environment, laser preparation of aluminum substrates is stable and shows higher tensile strength than other innovative treatments, such as atmospheric plasma, according to Rechner *et al.*

Studies on the durability of laser-treated adhesive-bonded titanium substrates were conducted by Kurtovic *et al* [36] and Rotella *et al* [35], using different aging conditions. In the first study, a clear correlation between surface morphology and durability of adhesive bonds, performing wedge tests, was established; it was proved that laser pre-treatment behaves much better than chemical methods. Also after aging in boiling water, as conducted in [35], results show that laser treatment improves joint strength before and after aging; the main effect individuated was the increased surface roughness and, consequently, the area available for bonding.

Other types of assessments on steel substrates have been performed by Rotella *et al* [3], who used a hydro-thermal cycle for a total period of 1000 h. A reduced loss of mechanical characteristics of bonded joints was observed, especially if compared to untreated or abrasion-treated joints.

5.3 Laser processing of polymeric materials to improve surface functionalities

Polymers present several advantages compared to metallic and ceramic materials, which make them an increasingly valuable choice in several industrial sectors, from electronics to biomedical applications, to the automotive and aerospace fields. The most exploited properties are: good specific mechanical resistance and toughness, high chemical inertness and good processability. The main drawback in some specific applications is the smooth surface characteristics of polymers, which cause surface modification and the creation of particular patterns at a nano-metric scale. In recent years, the interest on laser micro- and nano-processing has increased, due to the ability to fabricate microfluidic channels [42, 43], films for cell culture [44] or electronic devices [45].

One of the main problems of using lasers to modify surface wettability of polymers is that the absorptivity to laser radiation emitted by conventional laser sources varies greatly from material to material. Another issue is the low melting temperature of polymers, therefore in their laser processing the most critical parameter is pulse duration. In fact, efficiency and precision of the ablation mechanism is strongly linked to laser pulse duration. The ablation of polymers using long pulses could damage the material, creating an excess of molten materials and carbonization of the patterns' sides [46]. In contrast, ultra-short laser pulses (femtoseconds or picoseconds) bring limited thermal damage, which is beneficial in terms of texturing precision. In any case, as laser beam absorption varies for different materials, correct laser parameters must be selected for specific applications.

5.3.1 Influence of laser modification on surface energy and wettability characteristics

As reported above (section 5.2.1), modification of surface wettability concerns both chemical and morphological aspects. In the case of metals, the morphological aspect is more relevant, while the influence of surface chemistry varies greatly from metal to metal. In some cases it could be considered irrelevant. On the contrary, in the specific case of polymers, the formation of polar oxygen-containing groups, such as C–O, C=O, and O–C=O, are greatly responsible for the change in surface wettability [47]. An example is reported in [48], where the influence of different irradiations using a femtosecond laser source on PMMA substrates is studied. XPS was performed on untreated samples and specimens irradiated at a high fluence of 52.7 J cm^{-2} and irradiated at low fluence of 0.40 J cm^{-2}. Table 5.1 summarizes the results obtained.

At high fluence, a remarkable increase of O=C and O–C groups (polar groups) was detected. This phenomenon increases surface oxygen concentration and therefore hydrophilicity. On the contrary, at low fluence, there is a predominance of nonpolar groups (C–H and C–C) compared to that of the raw PMMA. An explanation of this behavior must be sought in the photodegradation caused by nonlinear absorption of the femtosecond laser, which is different from the thermal degradation associated with linear absorption in nanosecond pulses.

Similar effects were reported by Rebollar *et al* [44] using a nanosecond laser on PET films and investigated using Raman spectroscopy. In fact, an increase in the total C=O content is observed in the irradiated samples. A higher amount of oxygen atoms due to the appearance of new carboxylic and carbonyl groups was associated with a decrease in the CA value, suggesting that the surface shows more hydrophilic behavior after irradiation.

Table 5.1. Relative amount of oxygen/carbon functional groups (at.%) (reproduced from [48]).

Group	Control PMMA	High laser fluence	Low laser fluence
Nonpolar groups	54.09	31.16	60.23
Polar groups	45.90	68.84	39.78

5.3.2 Hydrophilicity and hydrophobicity of laser-treated surfaces

Obviously, keeping the definitions of hydrophilic and hydrophobic as seen in section 5.2.2 unchanged, for polymeric materials the wettability properties are mainly related to the type of laser source and to the specific energy supplied to the material itself, usually called fluence.

The laser is a valid alternative to the main traditional polymeric surface micro- and nano-scale modification methods: photolithography, electron or ion beam lithography, molding or chemical vapor deposition [49]. Nevertheless, traditional methods present several drawbacks: they do not comply to all the requirements needed for direct treatment of high performance polymeric materials [20]; they often require toxic chemical products or produce chemical waste; sterilization is not guaranteed (biomedical applications require a post-sterilization stage); and manufacturing of specific patterns is not always feasible.

Many studies report the application of laser surface texturing on several polymers, such as polyolefins, polyetheretherketone (PEEK), polycarbonate (PC) and polymethyl methacrylate (PMMA), for obtaining both hydrophobic and hydrophilic surfaces.

Riveiro *et al* [49] investigated surface texturing of polypropylene (PP) using nanosecond lasers at different wavelengths. Focusing on wettability, the 1064 nm and 355 nm laser wavelengths offer the best performance in terms of lower contact angles observed. This is probably linked to the small formation of C=O and hydroxyl groups on the surface of the laser-treated samples, inspected by surface chemical analyses. Similar results were obtained by Khaledian *et al* [50] for biomedical purposes. A decrease in CA of PP surfaces irradiated with UV laser radiation ($\lambda = 248$ nm) was related to the formation of polar functional groups. On another polyolefin, an ultra-high-molecular-weight polyethylene (UHMWPE) laser surface has been successfully used comparing different kinds of laser sources, such as femtosecond, picosecond and nanosecond [51].

On polycarbonate (PC) both hydrophilicity and hydrophobicity were studied as a function of the sources' wavelength. For example, Yilbas *et al* [52] report that a high power CO_2 laser (10.6 μm) could bring hydrophobic behavior to the treated surfaces. The same results were obtained using a nanosecond UV excimer laser (193 nm) [53] or using a nanosecond UV (263 nm) laser system [54]. Vice versa, nanosecond (ns) near-infrared Nd:YAG laser (1.064 μm) ablation can increase surface wettability after irradiation and bring hydrophilic behavior to the surface [55]. Recently, Riveiro *et al* [49] as well as Ravi-Kumar [46] have well summarized the state-of-the-art of wetting property modifications on several polymeric surfaces with laser surface irradiation.

5.3.3 Laser texturing to increase adhesive bonding of low surface energy substrates

Despite all of the advantages already cited, polymers present poor wettability, printability and low adhesion properties. Their increasing use in technical applications requires the need for effective joints to other components, and adhesive bonding is one of the most effective techniques. According to this perspective, tailoring of surface properties, such as wettability, chemical composition and texturing are often mandatory for successful applications. Traditional surface treatments are based both on chemical or physical modifications [56]. On one hand, mechanical abrasion is a typical physical method, which extends the bonding area increasing roughness, but could cause damage on the specimen surface and has poor repeatable results. On the other hand, several chemical treatments have been developed with the aim of modifying both the morphology and chemical structure of polymers, but they present environmental issues concerning waste disposal.

Although the effects on surface characteristics resulting from laser radiation on polymers appear to be useful for adhesive bonding applications, the literature is not extensive on this aspect [55, 57–59].

One of the first studies was conducted by Laurens *et al* [59] on PEEK films, using a pulsed excimer laser. The results have shown an improvement in adhesive bonding properties of laser-treated substrates, which was related by the authors to the increased wettability and the introduction of polar species on the surfaces.

Similar results on PEEK, using a different laser source, were found by Wilson *et al* [55], who report the effects of surface treatment with a nanosecond pulsed Nd:YAG laser in the near-infra-red region, at different laser fluences. Single lap-shear testing of PEEK joints showed great improvement of the shear strength of the laser-treated adhesively-bonded joints, reaching a strength 13 times higher than untreated PEEK.

Mandolfino *et al* [58] investigated the effect of different laser pre-treatments on polyolefinic substrates using both IR and UV laser sources. The results showed a substantial increase in mechanical resistance of single lap-bonded joints for high fluence values.

The excellent potentialities of laser texturing in surface treatment of substrates in order to increase the resistance of polymeric-bonded joints are therefore evident.

5.3.4 Changes in durability behavior of adhesive-bonded joints

As mentioned above, the durability of joints is a particularly critical issue, because it is one of the major shortcomings of adhesion bonding, a polymeric adhesive being employed in the vast majority of cases. In the case of joints made also with polymeric substrates the problem becomes even more critical, since they are particularly subject to deterioration due to moisture, UV rays, heat, etc [60].

Many authors have studied the durability of the adhesive itself [61–64] or the adhesive-bonded samples, in several environments and using several polymeric substrates [65–69]. Many previous studies are well summarized by Ebnesajjad [41].

Only a small number of studies highlighted the effect of surface pre-treatment on the durability of bonded joints and compared different treatments [70, 71]. To the best of the authors' knowledge, the effect of laser surface treatment to increase the durability of polymer-bonded joints has not been studied yet in depth and remains a crucial point that is open to further investigation.

5.4 Laser processing of CFRP substrates to improve surface functionalities

Carbon fiber composites, particularly those with a polymeric matrix, demonstrate many properties of significant interest for many technical applications. High specific mechanical strength, modulus, temperature performance and corrosion resistance are the most exploited.

The sectors particularly interested in the use of this material are aerospace, automotive, sporting goods, construction, biomedical and microelectronics, among others. Most of the applications demand high surface quality and precision in machining the material [72].

Most of the studies that investigate surface functionalities are concerned with the improvement of adhesive properties, both for the realization of homogeneous and heterogeneous joints, especially with metals.

5.4.1 Influence of laser modification on surface energy and wettability characteristics

The study of the wettability of CFRP substrates was mainly carried out in relation to its application to adhesive-bonded joints. One of the most crucial steps before performing CFRP adhesive bonding is surface pre-treatment, because very often the polymer matrix exhibits low surface energy and wettability. Many surface treatments have been developed for this type of substrate and have been studied, such as: solvent cleaning, mechanical abrasion [73, 74], peel-ply [75], chemical etching [74] or plasma treatments, both low pressure [76] or atmospheric plasma [77]. The purpose of so-called physical treatments, i.e. solvent cleaning or mechanical abrasion, is the elimination of surface contaminants and the roughening of substrates to increase the real bonding area. These treatments have proved to be more effective for thermosetting matrix composites than for those with a thermoplastic matrix, which have a lower wettability, as summarized by Molitor *et al* [78]. Many studies focused on physical and chemical procedures, such as plasma, with the aim of functionalizing the specimens combining several effects, which are mainly cleaning, activation and inserting polar species on polymeric surfaces without affecting bulk properties [79–82].

In parallel, laser surface pre-treatment of composite materials has been the subject of many studies, which, however, are more concentrated on the morphological aspects of surfaces and the mechanical characteristics of bonded joints, rather than on surface wettability [83–86]. The most interesting and recent works in this regard are by Sun *et al* [87] and Akman *et al* [88].

Sun *et al* evaluated the wetting performance of a laser-treated CFRP substrate surface (using an energy density of 35.4 mJ mm^{-2}) and report an improvement with smaller CAs than untreated CFRP substrates' surface. This was related to the increase of polar components, because of the removal of the release agent, which is an inert substance with low surface tension, and a partial oxidation of the carbon fibers' surface, which facilitates the mutual adsorption of carbon fibers and polar molecules [87].

Recently, Akman *et al* evaluated the effects of accumulated laser fluence as surface treatment of CFRP using pulsed CO_2. The results demonstrate a CA value dependent on laser fluence and the amount of matrix on the surface. In fact, water CA exhibited a symmetrical distribution for low accumulated laser fluences (<18 J cm^{-2}). By decreasing the amount of epoxy matrix on the surface due to the high fluences laser treatment (>18 J cm^{-2}), the water droplet begins to spread in the fiber direction, while in the other axis it remains approximately constant. This change in CA behavior can be explained by the removal of the polymeric matrix around the fibers and the water droplet tendency to fill these volumes. Figure 5.6 compares contact angles and surface roughness of differently treated CFRP samples.

Although there are not many studies that focus only on surface wettability aspects, most confirm that good wettability can be achieved and that this feature is linked to an increase in the mechanical properties of the bonded joints.

5.4.2 Laser texturing to increase adhesive bonding of low surface energy substrates

Mechanical joining methods, such as rivets, bolts or flow drill screws, do not allow one to fully exploit the lightweight potential of CFRP, because they break the fibers and create notches that cause stress concentrations in proximity of the holes. Moreover, metal fasteners could significantly increase the weight of the whole structure.

Adhesive bonding is therefore an optimal alternative to overcome these drawbacks. In fact, it does not require any drilling or hole in the CFRP, it distributes

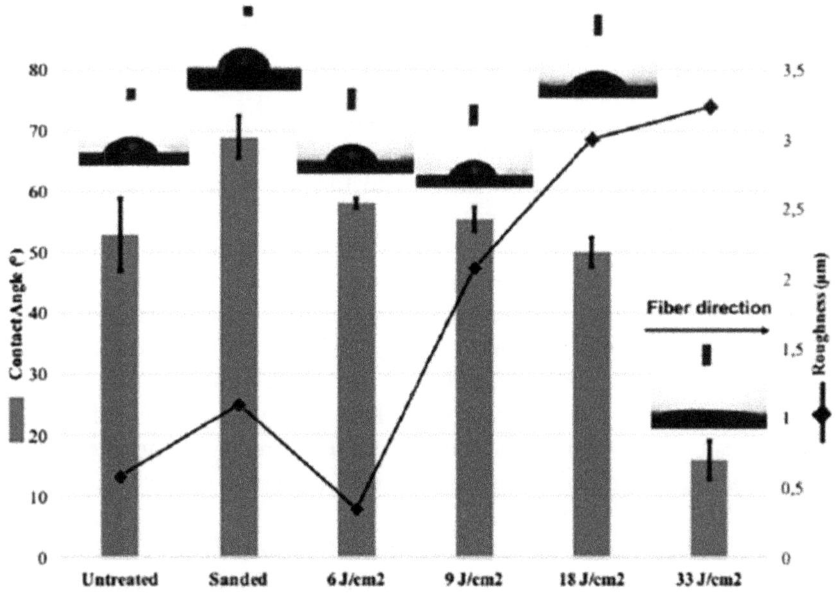

Figure 5.6. Contact angles and roughness of untreated, abraded and laser treated at different fluences' samples. Reprinted from [88], copyright (2019), with permission from Elsevier.

mechanical stress over the entire bonded area and, if a polymeric adhesive is used, it does not increase the final assembly weight [84].

As mentioned above, in order to realize a high performance adhesive bond for other kinds of substrates, it is necessary to properly prepare the surface by removing any contaminants, i.e. mold release agents and other contaminant particles from previous manufacturing steps. Of course, this surface cleaning must be realized without damaging the fibers. The traditional surface preparations were mainly mechanical abrasion and peel-ply, but they present several disadvantages.

In fact, the automation of an abrasion method can be quite difficult, and solvent cleaning is typically necessary both before and after this step. For this reason, this step is usually performed manually, which makes it unsuitable for mass production and large CFRP substrate areas. Most importantly, manual treatment significantly reduces process repeatability, possibly leading to two opposite scenarios: fiber damage because of a too aggressive abrasion, or weak contaminants removal [84].

Also, the use of peel plies is not free from technical issues. They can be difficult to remove, especially as during the high temperature CFRP cure cycle these tend to bind to the matrix of the composite itself. Kanerva and Saarela observed that also in terms of mechanical behavior, the strength of a CFRP joint realized using peel plies is rarely higher than a joint pre-treated by mechanical abrasion. The roughness left by tearing a peel-ply off a CFRP surface is full of ridges and bumps and they can disadvantageously affect the bond strength. The explanation is probably that air could get trapped in the roughness peaks and valleys forming voids. Consequently, alternative surface treatments, such as plasma and laser, have been studied, especially for automated production [75].

In the case of polymeric matrix composite materials, one of the main advantages of laser sources is the possibility of adapting parameters to be able to remove material on the surface from less than a micron up to tens of microns, calibrating in particular the frequency pulses. Furthermore, a laser is highly selective and, in this way, the unclean matrix resin can be removed from a composite surface without damaging fibers or avoiding the complete exposure of fibers on the surface. The effect of contamination removal is similar to mechanical abrasion, but, since no residual grit is produced, no subsequent cleaning is required. In this way, laser surface treatment is much more repeatable and could be easily automatized to prepare small to large CFRP substrates surfaces for bonding [89].

In order to achieve highly controlled and repeatable surface properties, the laser parameters to be taken into account are mainly: power, pulse frequency, spot size and travel speed. Usually, the optimal state for CFRP surface preparation is the ablation. To reach the ablation threshold, a laser pulse must have sufficient fluence (i.e. energy density) and bring an absorbent material to complete vaporization. Of course, the ablation threshold is strongly related to the material (i.e. absorbing coefficient) and to the laser process.

Successful experiences of laser surface preparation of CFRP prior to bonding are reported in several works [83, 85–87, 90–93], using different laser sources and different wavelengths.

Zhan *et al* investigated the photochemical ablation reaction on epoxy resin samples irradiated with a UV laser, studying the effect of different resin removal ratios on the shear strength of the bonded single lap joint. The most interesting finding was that the surface with a small resin amount gave better results, in terms of adhesive bonding strength, compared to the samples with no resin and carbon fibers exposed [93].

Similar results were obtained by Sun *et al*, who optimized the parameters of an IR fiber laser and found that the ablation treatment successfully increases the lap-shear strength of CFRP single lap joints. In particular, the best energy density value was identified and an efficient contaminants and release agent removal, without damaging the carbon fibers, was obtained, with the final effect of increasing the strength of CFRP adhesive-bonded joints [87].

Bora *et al* instead focused their study on the generation of microholes on the surface of a unidirectional carbon fiber-reinforced epoxy composite, creating optimal conditions for mechanical interlocking. Different patterns were created and compared based on their effectiveness on the increase of the adhesive bond strength. The best results were obtained with a frame-type pattern, which showed a strength increase of about 30%. [83].

5.4.3 Changes in durability behavior of adhesive-bonded joints

Temperature and humidity are commonly considered to be the main factors that could affect the adhesive bonding performance of CFRP composite systems [74, 94–96].

Concerning traditional pre-treatment, environmental effects on the mechanical performance of CFRP adhesive-bonded joints have been studied rather widely. The effects of peel-ply were summarized by Kanerva and Saarela [75].

On the contrary, the effects of aging on the mechanical performance of laser pre-treated joints have not been studied extensively. Palmieri *et al* hygrothermally [97] aged adhesive-bonded joints in an environmental chamber in controlled conditions (71°C – 85% relative humidity) for three years and tested some of them periodically, to measure apparent shear strengths and determine failure modes.

5.5 Conclusion

Laser-induced surface texturing at the micro- and nano-scale is a new technology in the service of surface engineering, with the possibility of creating very fine structures with extremely high resolution. A wide range of materials can be processed, from metal alloys to polymers and composites, both in laboratory and industrial applications.

For all kinds of substrates, it was observed that the choice of the most suitable laser source in reference to the material to be treated is critical, as well as the optimization of parameters to obtain the desired effect, considering both short and long term applications.

References

[1] Ion J C 2005 Introduction *Laser Processing of Engineering Materials: Principles and Industrial Application* ed J C Ion (London: Butterworth)

[2] Ion J C 2005 Lasers *Laser Processing of Engineering Materials: Principles and Industrial Application* ed J C Ion (London: Butterworth) pp 41–103

[3] Ion J C 2007 Evolution of laser material processing *Laser Process. Eng. Mater.* **1505** 12–40

[4] Chikarakara E, Naher S and Brabazon D 2012 High speed laser surface modification of Ti–6Al–4V *Surf. Coatings Technol.* **206** 3223–29

[5] Ion J C 2005 Surface melting *Laser Processing of Engineering Materials: Principles and Industrial Application* ed J C Ion (London: Butterworth) pp 261–95

[6] Elliott T L 1978 Surface hardening *Tribol. Int.* **11** 121–25

[7] Łęcka K M, Gąsiorek J, Mazur-Nowacka A, Szczygieł B and Antończak A J 2018 Adhesion and corrosion resistance of laser-oxidized titanium in potential biomedical application *Surf. Coatings Technol.* **366** 179–89

[8] Davis J R 2001 Introduction to surface engineering for corrosion and wear resistance *Surface Engineering for Corrosion and Wear Resistance* (Novelty, OH: ASM International) pp 1–10

[9] Ashby M F and Jones D R H 2013 Processing metals 2 *Eng. Mater* **2** 279–96

[10] Fuentes G G 2015 Surface engineering and micro-manufacturing *Micromanufacturing Engineering and Technology* 2nd edn ed Y Qin (Boston, MA: William Andrew) pp 459–86

[11] Martin P M 2010 Deposition technologies: an overview *Handbook of Deposition Technologies for Films and Coatings: Science, Applications and Technology* (Boston, MA: William Andrew Publishing) pp 1–31

[12] Bizi-bandoki P, Benayoun S, Valette S, Beaugiraud B and Audouard E 2011 Modifications of roughness and wettability properties of metals induced by femtosecond laser treatment *Appl. Surf. Sci.* **257** 5213–18

[13] Allahyari E *et al* 2018 Laser surface texturing of copper and variation of the wetting response with the laser pulse fluence *Appl. Surf. Sci.* **470** 817–24

[14] Li J, Zhou Y, Fan F, Du F and Yu H 2019 Controlling surface wettability and adhesive properties by laser marking approach *Opt. Laser Technol.* **115** 160–65

[15] Sciancalepore C, Gemini L, Romoli L and Bondioli F 2018 Study of the wettability behavior of stainless steel surfaces after ultrafast laser texturing *Surf. Coat. Technol.* **352** 370–77

[16] Wang X, Zheng H, Wan Y, Feng W and Lam Y C 2018 Picosecond laser surface texturing of a stavax steel substrate for wettability control *Engineering* **4** 816–21

[17] Bizi-Bandoki P, Valette S, Audouard E and Benayoun S 2013 Time dependency of the hydrophilicity and hydrophobicity of metallic alloys subjected to femtosecond laser irradiations *Appl. Surf. Sci.* **273** 399–407

[18] Kietzig A M, Hatzikiriakos S G and Englezos P 2009 Patterned superhydrophobic metallic surfaces *Langmuir* **25** 4821–27

[19] Cassie B D 1944 Wettability of porous surfaces *Trans. Faraday Soc.* **40** 546–51

[20] Wenzel R N 1936 Resistance of solid surfaces to wetting by water *Ind. Eng. Chem.* **28** 988–94

[21] Gu Y *et al* 2017 Research progress of biomimetic superhydrophobic surface characteristics, fabrication, and application *Adv. Mech. Eng.* **9** 1–13

[22] Ahmmed K M T, Ling E J Y, Servio P and Kietzig A M 2015 Introducing a new optimization tool for femtosecond laser-induced surface texturing on titanium, stainless steel, aluminum and copper *Opt. Lasers Eng.* **66** 258–68

[23] Grabowski A, Sozańska M, Adamiak M, Kępińska M and Florian T 2018 Laser surface texturing of Ti6Al4V alloy, stainless steel and aluminium silicon alloy *Appl. Surf. Sci.* **461** 117–23

[24] Ahmad D, van den Boogaert I, Miller J, Presswell R and Jouhara H 2018 Hydrophilic and hydrophobic materials and their applications *Energy Sources, Part A Recover. Util. Environ. Effects* **40** 2686–725

[25] Gao L and Mccarthy T J 2008 Teflon is hydrophilic. Comments on Definitions of hydrophobic, shear versus tensile hydrophobicity, and wettability characterization *Langmuir* **24** 546–50

[26] Drelich J, Chibowski E, Meng D D and Terpilowski K 2011 Hydrophilic and super-hydrophilic surfaces and materials *Soft Matter* **7** 9804–28

[27] Zhang X, Shi F, Niu J, Jiang Y and Wang Z 2008 Superhydrophobic surfaces: from structural control to functional application *J. Mater. Chem.* **18** 621–33

[28] Tavana H, Amirfazli A and Neumann A W 2006 Fabrication of superhydrophobic surfaces of *n*-hexatriacontane *Langmuir* **22** 5556–59

[29] Zhao J *et al* 2019 A rapid one-step nanosecond laser process for fabrication of super-hydrophilic aluminum surface *Opt. Laser Technol.* **117** 134–41

[30] Fleming R A and Zou M 2014 Fabrication of stable superhydrophilic surfaces on titanium substrates *J. Adhes. Sci. Technol.* **28** 823–32

[31] Wu Y *et al* 2016 Effect of laser ablation surface treatment on performance of adhesive-bonded aluminum alloys *Surf. Coatings Technol.* **304** 340–47

[32] Zhu C *et al* 2019 Application of pulsed Yb: fiber laser to surface treatment of Al alloys for improved adhesive bonded performance *Opt. Lasers Eng.* **119** 65–76

[33] Romoli L, Moroni F and Khan M M A 2017 A study on the influence of surface laser texturing on the adhesive strength of bonded joints in aluminium alloys *CIRP Ann. Manuf. Technol.* **66** 237–40

[34] Mandolfino C, Lertora E, Genna S, Leone C and Gambaro C 2015 Effect of laser and plasma surface cleaning on mechanical properties of adhesive bonded joints *Procedia CIRP* **33** 458–63

[35] Rotella G, Orazi L, Alfano M, Candamano S and Gnilitskyi I 2017 Innovative high-speed femtosecond laser nano-patterning for improved adhesive bonding of Ti6Al4V titanium alloy *CIRP J. Manuf. Sci. Technol.* **18** 101–6

[36] Kurtovic A, Brandl E, Mertens T and Maier H J 2013 Laser induced surface nano-structuring of Ti–6Al–4V for adhesive bonding *Int. J. Adhes. Adhes.* **45** 112–7

[37] Lunder O, Lapique F, Johnsen B and Nisancioglu K 2004 Effect of pre-treatment on the durability of epoxy-bonded AA6060 aluminium joints *Int. J. Adhes. Adhes.* **24** 107–17

[38] Brack N and Rider A N 2014 The influence of mechanical and chemical treatments on the environmental resistance of epoxy adhesive bonds to titanium *Int. J. Adhes. Adhes.* **48** 20–7

[39] Rotella G, Alfano M, Schiefer T and Jansen I 2015 Enhancement of static strength and long term durability of steel/epoxy joints through a fiber laser surface pre-treatment *Int. J. Adhes. Adhes.* **63** 87–95

[40] Rechner R, Jansen I and Beyer E 2010 Influence on the strength and aging resistance of aluminium joints by laser pre-treatment and surface modification *Int. J. Adhes. Adhes.* **30** 595–601

[41] Ebnesajjad S 2009 Durability of adhesive bonds *Adhesives Technology Handbook* (Boston, MA: William Andrew) pp 231–72

[42] Teixidor D, Ciurana J, Thepsonthi T and Özel T 2012 Nanosecond pulsed laser micromachining of PMMA-based microfluidic channels *Trans. North Am. Manuf. Res. Inst. SME* **40** 536–43

[43] Chang T L, Tsai T K, Yang H P and Huang J Z 2012 Effect of ultra-fast laser texturing on surface wettability of microfluidic channels *Microelectron. Eng.* **98** 684–88

[44] Rebollar E, Castillejo M and Ezquerra T A 2015 Laser induced periodic surface structures on polymer films: from fundamentals to applications *Eur. Polym. J.* **73** 162–74

[45] Chen Z and Webster D C 2006 Study of cationic UV curing and UV laser ablation behavior of coatings sensitized by novel sensitizers *Polymer (Guildf)* **47** 3715–26

[46] Ravi-Kumar S, Lies B, Zhang X, Lyu H and Qin H 2019 Laser ablation of polymers: a review *Polym. Int.* **34** 316–27

[47] Vesel A and Mozetič M 2015 Low-pressure plasma-assisted polymer surface modifications *Printing on Polymers: Fundamentals and Applications* ed J Izdebska and S Thomas (Boston, MA: William Andrew) pp 101–21

[48] Zheng H Y, Guan Y C, Liu K, Wang Z K and Yuan S M 2016 Other methods of polymer surface modifications *Printing on Polymers: Fundamentals and Applications* ed J Izdebska and S Thomas (Boston, MA: William Andrew Publishing) pp 161–78

[49] Riveiro A, Maçon A L B, del Val J, Comesaña R and Pou J 2018 Laser surface texturing of polymers for biomedical applications *Front. Phys.* **5** 16

[50] Khaledian M, Jiroudhashemi F and Biazar E 2017 Chitosan- and polypropylene-oriented surface modification using excimer laser and their biocompatibility study *Artif. Cells, Nanomed. Biotechnol.* **45** 135–38

[51] Riveiro A *et al* 2014 Laser surface modification of ultra-high-molecular-weight polyethylene (UHMWPE) for biomedical applications *Appl. Surf. Sci.* **302** 236–42

[52] Yilbas B S *et al* 2014 Wetting and other physical characteristics of polycarbonate surface textured using laser ablation *Appl. Surf. Sci.* **320** 21–9

[53] Zenkiewicz M, Rytlewski P, Tracz A, Mróz W and Richert J 2009 Effects of laser radiation on some properties of the surface layer of polycarbonate *Polimery* **54** 639–47

[54] Lasagni A F, Alamri S, Aguilar-Morales A I, Rößler F, Voisiat B and Kunze T 2018 Biomimetic surface structuring using laser based interferometric methods *Appl. Sci.* **8** 1260

[55] Wilson A, Jones I, Salamat-Zadeh F and Watts J F 2015 Laser surface modification of poly (etheretherketone) to enhance surface free energy, wettability and adhesion *Int. J. Adhes. Adhes.* **62** 69–77

[56] ASTM International 2003 *Standard Practice for Preparation of Surfaces of Plastics Prior to Adhesive. ASTM D2093-03* (West Conshohocken, PA: ASTM International)

[57] Buchman A, Dodiuk H, Rotal M and Zahavi J 1991 Preadhesion treatment of thermoplastic adherends using excimer laser *Int. J. Adhes. Adhes.* **11** 144–49

[58] Mandolfino C, Pizzorni M, Lertora E and Gambaro C 2019 Laser surface pre-treatment of polyolefin substrates for adhesive bonding *Proc. of the 22nd Int. ESAFORM Conf. on Material Forming: ESAFORM 2019 (Vitoria-Gasteiz, Spain, 8–10 May 2019) AIP Conf. Proc.* 2113

[59] Laurens P, Sadras B, Decobert F and Amouroux J 1998 Enhancement of the adhesive bonding properties of PEEK by excimer laser treatment *Int. J. Adhes. Adhes.* **18** 19–27

[60] Badia J D, Gil-Castell O and Ribes-Greus A 2017 Long-term properties and end-of-life of polymers from renewable resources *Polym. Degrad. Stab.* **137** 35–57

[61] Del Real J C, De Santayana M C, Abenojar J and Martinez M A 2006 Adhesive bonding of aluminium with structural acrylic adhesives: durability in wet environments *J. Adhes. Sci. Technol.* **20** 1801–18

[62] Chamochin R, De Santayana M C, Abenojar J, Pantoja M, Ballesteros Y and Del Real J C 2010 The effect of surface treatment on the behavior of toughened acrylic adhesive/GRP (epoxy) composite joints *J. Adhes. Sci. Technol.* **24** 1903–16

[63] Comyn J 2012 Durability of adhesives in wet conditions *Adhesives in Marine Engineering* (Cambridge: Woodhead Publishing) pp 187–207

[64] Sousa J M, Correia J R and Cabral-Fonseca S 2018 Durability of an epoxy adhesive used in civil structural applications *Constr. Build. Mater.* **161** 618–33

[65] Goss B 2010 Durability and environmental testing *Practical Guide to Adhesive Bonding of Small Engineering Plastic and Rubber Parts Practical Guide Series* (Shrewsbury: iSmithers Rapra Publishing) pp 127–37

[66] Gao Z, Peng S, Sun J, Yao L and Qiu Y 2010 The influence of moisture on atmospheric pressure plasma etching of PA6 films *Curr. Appl Phys.* **10** 230–34

[67] Gao Z, Peng S, Sun J, Yao L and Qiu Y 2009 Influence of processing parameters on atmospheric pressure plasma etching of polyamide 6 films *Appl. Surf. Sci.* **255** 7683–88

[68] Enciso B, Abenojar J and Martínez M A 2017 Influence of plasma treatment on the adhesion between a polymeric matrix and natural fibres *Cellulose* **24** 1791–801

[69] Encinas N, Díaz-Benito B, Abenojar J and Martínez M A 2010 Extreme durability of wettability changes on polyolefin surfaces by atmospheric pressure plasma torch *Surf. Coatings Technol.* **205** 396–402

[70] Mandolfino C, Lertora E, Gambaro C and Pizzorni M 2018 Durability of polyamide bonded joints: influence of surface pre-treatment *Int. J. Adhes. Adhes.* **86** 123–30

[71] Critchlow G W, Lee R J, Hutchinson A R and Wingfield J R J 1993 MTS Project 4, Characterisation of Surface Condition, Report No. 2: Review of Substrate Surface Treatments, September 1993, Report carried out by AEA Technology in collaboration with Oxford Brookes, Loughborough and Surrey Universities http://www.adhesivestoolkit.com/PDFFiles/Project%204/P4r2.pdf

[72] Strong A B 2008 Composites applications, *Fundamentals of Composites Manufacturing: Materials, Methods, and Applications* 2nd edn; A B Strong (Southfield, MI: Society of Manufacturing Engineers) pp 543–78

[73] Baldan A 2004 Review Adhesively-bonded joints and repairs in metallic alloys, polymers and composite materials: adhesives, adhesion theories and surface pretreatment *J. Mater. Sci.* **39** 1–49

[74] 2002 *Department of Defense Handbook: Composite Materials Handbook - Polymer Matrix Composites Guidelines for Characterization of Structural Materials* vol 1 MIL-HDBK-17/1F, USA http://everyspec.com/MIL-HDBK/MIL-HDBK-0001-0099/download.php?spec=MIL_HDBK_17_1F.237.pdf

[75] Kanerva M and Saarela O 2013 The peel ply surface treatment for adhesive bonding of composites: a review *Int. J. Adhes. Adhes.* **43** 60–9

[76] Pizzorni M, Lertora E, Gambaro C, Mandolfino C, Salerno M and Prato M 2019 Low-pressure plasma treatment of CFRP substrates for epoxy-adhesive bonding: an investigation of the effect of various process gases *Int. J. Adv. Manuf. Technol.* **102** 3021–35

[77] Dighton C, Rezai A, Ogin S L and Watts J F 2019 Atmospheric plasma treatment of CFRP composites to enhance structural bonding investigated using surface analytical techniques *Int. J. Adhes. Adhes.* **91** 142–49

[78] Molitor P, Barron V and Young T 2001 Surface treatment of titanium for adhesive bonding to polymer composites: a review *Int. J. Adhes. Adhes.* **21** 129–36

[79] Encinas N *et al* 2014 Surface modification of aircraft used composites for adhesive bonding *Int. J. Adhes. Adhes.* **50** 157–63

[80] Lee H, Ohsawa I and Takahashi J 2015 Effect of plasma surface treatment of recycled carbon fiber on carbon fiber-reinforced plastics (CFRP) interfacial properties *Appl. Surf. Sci.* **328** 241–46

[81] Rhee K Y and Yang J H 2003 A study on the peel and shear strength of aluminum/CFRP composites surface-treated by plasma and ion assisted reaction method *Compos. Sci. Technol.* **63** 33–40

[82] Baldan A 2004 Adhesively-bonded joints and repairs in metallic alloys, polymers and composite materials: adhesives, adhesion theories and surface pretreatment *J. Mater. Sci.* **39** 1–49

[83] Bora M Ö, Akman E, Çoban O, Genc Oztoprak B and Demir A 2019 The effect of CO_2 laser-induced microhole formations on adhesive bonding strength of CFRP/CFRP joints *Polym. Compos.* **40** 2891–900

[84] Fischer F, Kreling S, Gäbler F and Delmdahl R 2013 Using excimer lasers to clean CFRP prior to adhesive bonding *Reinf. Plast.* **57** 43–6

[85] Loutas T H, Sotiriadis G, Tsonos E, Psarras S and Kostopoulos V 2019 Investigation of a pulsed laser ablation process for bonded repair purposes of CFRP composites via peel testing and a design-of-experiments approach *Int. J. Adhes. Adhes.* **95** 102407

[86] Moreira R D F, Oliveira V, Silva F G A, Vilar R and de Moura M F S F 2018 Influence of femtosecond laser treated surfaces on the mode I fracture toughness of carbon-epoxy bonded joints *Int. J. Adhes. Adhes.* **82** 108–13

[87] Sun C, Min J, Lin J, Wan H, Yang S and Wang S 2018 The effect of laser ablation treatment on the chemistry, morphology and bonding strength of CFRP joints *Int. J. Adhes. Adhes.* **84** 325–34

[88] Akman E, Erdoğan Y, Bora M Ö, Çoban O, Oztoprak B G and Demir A 2019 Investigation of accumulated laser fluence and bondline thickness effects on adhesive joint performance of CFRP composites *Int. J. Adhes. Adhes.* **89** 109–16

[89] Palmieri F L *et al* 2019 Optimized surface treatment of aerospace composites using a picosecond laser *Compos. Part B Eng.* **175** 107155

[90] Belcher M A, Wohl C J and Connell J W 2009 Laser surface preparation and bonding of aerospace structural composites *Proc. of the 17th Int. Conf. on Composite Materials (Edinburgh)* Paper ID: IF16.1

[91] Palmieri F L, Belcher M A, Wohl C J, Blohowiak K Y and Connell J W 2016 Laser ablation surface preparation for adhesive bonding of carbon fiber reinforced epoxy composites *Int. J. Adhes. Adhes.* **68** 95–101

[92] Zhan X, Li Y, Gao C, Wang H and Yang Y 2018 Effect of infrared laser surface treatment on the microstructure and properties of adhesively CFRP bonded joints *Opt. Laser Technol.* **106** 398–409

[93] Zhan X, Chen S, Li Y, Wang H and Yang Y 2019 Effect of surface cold ablation on shear strength of CFRP adhesively bonded joint after UV laser treatment *Int. J. Adhes. Adhes.* **94** 13–23

[94] Parker B M 1983 The effect of composite prebond moisture on adhesive-bonded CFRP–CFRP joints *Composites* **14** 226–32

[95] John S J, Kinloch A J and Matthews F L 1991 Measuring and predicting the durability of bonded carbon fibre/epoxy composite joints *Composites* **22** 121–27
[96] Parker B M 1990 The strength of bonded carbon fibre composite joints exposed to high humidity *Int. J. Adhes. Adhes.* **10** 187–91
[97] Palmieri F L, Belcher M A, Wohl C J, Blohowiak K Y and Connell J W 2016 Laser ablation surface preparation for adhesive bonding of carbon fiber reinforced epoxy composites *Int. J. Adhes. Adhes.* **68** 95–101

IOP Publishing

Laser Micro- and Nano-Scale Processing
Fundamentals and applications
Ahmed Issa and Dermot Brabazon

Chapter 6

Investigation methods to understand laser-induced surface modification

Chiara Mandolfino, Silvia Vicini, Maila Castellano, Enrico Lertora, Marco Pizzorni and Carla Gambaro

6.1 Introduction

Laser micro- and nano-texturing is becoming an increasingly important tool in the field of advanced surface engineering, which is a particularly relevant sector for several manufacturing processes. The modification introduced on the surface through the interaction of a laser beam with different kinds of materials, requires the application of advanced characterization techniques, that can highlight aspects useful for the industrial application of the modifications themselves.

In fact, a technical background on surface characterization is necessary to better answer surface-related problems during advanced manufacturing. In the specific case of laser processing used as surface pre-treatment, the related characterization techniques focus on wettability, surface topography and chemical composition.

The aim of this chapter is to provide a general and concise overview of surface characterization techniques, as it is useful to understand the effect of laser surface texturing both on the micro- and nano-scale. In the first section, wettability theories will be described and their link to the results of experimental investigation on laser surface texturing will be presented.

Since laser processing, superficial morphology is one of the most easily modifiable characteristics, and the following section will focus on methods of investigation that allow qualitative and quantitative detection of these changes. Finally, investigations concerning the possible chemical modification, due to oxidation states or any type of interaction of the surface layers with the surrounding environment, will be reported.

6.2 Wettability of surfaces

Wettability of solid surfaces involves many research and practical applications: adhesion, wear and corrosion resistance, bio-compatibility and antifouling are the

doi:10.1088/978-0-7503-1683-5ch6

most representative ones [1–5]. In particular, the possibility of tailoring wettability is strongly related to the surface preparation, which must be designed specifically for each material.

In an ideal situation, once a drop is placed on a horizontal smooth surface, it could form a sphere or spread completely. Considering the drop, the angle formed at the interface of the different phases, i.e. solid, liquid and gas at equilibrium, is called the contact angle (figure 6.1).

The equilibrium is reached at the minimal energy state among the three phases. The theory was first described by the Young relation [7], which relates the contact angle and surface free energy (SFE) of a solid.

The contact angle is one of the most important methods to quantify the wetting of liquid–solid contact. The Young relation is valid for an ideal surface, but considering the real surface, other factors must be taken into account: surface condition and surface roughness are the most important ones in tuning and controlling the wettability [8].

The first effort to describe the wettability of a rough surface was made by Wenzel [9]. According to this theory, the real surface, which presents a certain roughness, enlarges the solid–liquid interface area in comparison to the ideal smooth surface. The relationship describing this situation is as follows:

$$\cos \theta_a = r \cos \theta$$

where θ_a is the apparent CA for a rough surface, θ represents the CA for a smooth surface and r is the ratio of the real surface area to its flat projected area, which define theoretically the roughness of the material. According to practical experience, this model is useful in the description of the contact angle for simple morphologies and roughness trends and for a wetting surface (range $0° < \theta < 90°$). By increasing the contact angle values (range $90° < \theta < 180°$), it has been found that the liquid could not penetrate well in the surface asperities and gas molecules can be trapped in the roughness valleys. In this case, a discontinuity at the interface between liquid and solid is generated [2]. Cassie [10] proposed another equation for heterogeneous

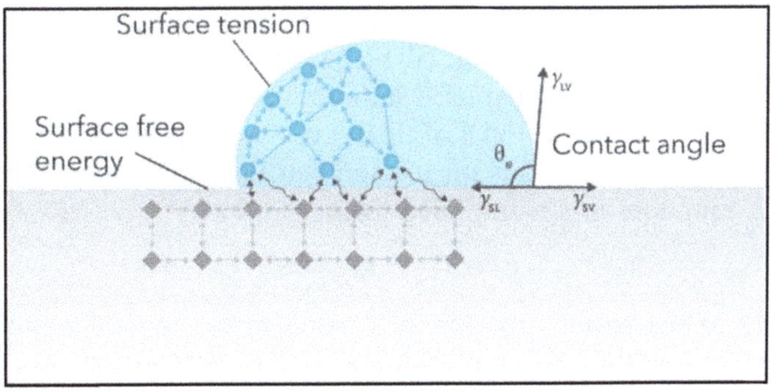

Figure 6.1. Contact angles formed by sessile liquid drops on an ideal solid surface, surface tension in the liquid and surface free energy (SFE) of the interface surface. Courtesy of Nanoscience Instruments [6].

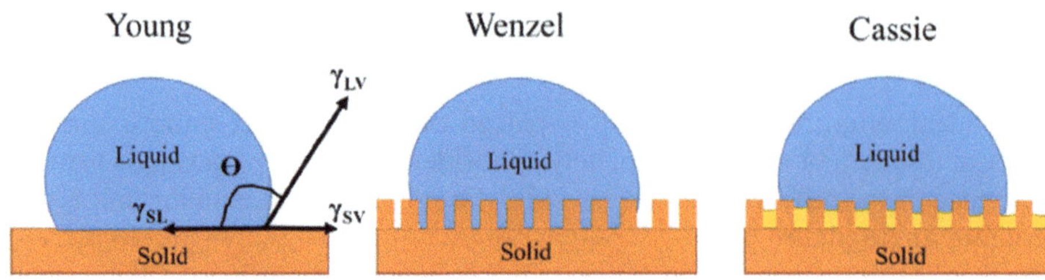

Figure 6.2. Different models of wetting of solid surfaces. Reprinted from [8], copyright (2018), with permisson from Elsevier.

surfaces (alternation of liquid, solid and gas at the interface) basing the study on composite materials:

$$\cos \vartheta_a = s_1(\cos \vartheta + 1) - 1$$

where s_1 is the area fraction of the solid surface in contact with the liquid droplet. This theory could be applied to rough surfaces, including the effect of trapped air or gas in the roughness asperities.

Figure 6.2 compares the different wetting models discussed above.

More recently, different authors proposed an elaboration of the classic models in order to describe real surfaces and a correlation between roughness parameters (in 2D and 3D, as defined by ISO 25178 standards) and contact angles [2, 11–13].

6.2.1 Mathematical models to relate wettability and surface energy

Laser surface texturing is used more and more to modify adhesion properties, for adhesive bonding, painting but also self-cleaning surface applications. SFE is one of the most used parameters to quantify adhesion properties [14].

SFE is a physical phenomenon, which describes the equilibrium of atoms in the surface topmost layer of materials, caused by intermolecular interactions at an interface between two mediums. This energy is different for each material and this term is used specifically for solid surfaces. The unit of SFE is J m^{-2}. When referring to liquids, the same physical quantity is described by the term surface tension γ (unit N m^{-1}).

Usually the surface tension of liquid could be easily determined using direct methods [15, 16], while measuring the SFE of solids could be more problematic and no direct methods exist. The most used indirect methods are those of Fowkes [17], Owens–Wendt [18] and Wu [19]. All of them are based on the evaluation of contact angles of two or more test liquids on the solid surface, and calculation of the SFE using Young's equation, in its simplified form, is

$$\gamma_S = \gamma_{SL} + \gamma_L \cdot \cos \theta$$

where γ_S refers to the surface tension of a solid in a vacuum medium, γ_{SL} is the surface tension at the solid–liquid interface, γ_L is the surface tension of the measured liquid and θ is the contact angle at equilibrium state.

In order to obtain the value of SFE according to these methods, it is necessary to use various liquids with known characteristics. The most used are deionised water,

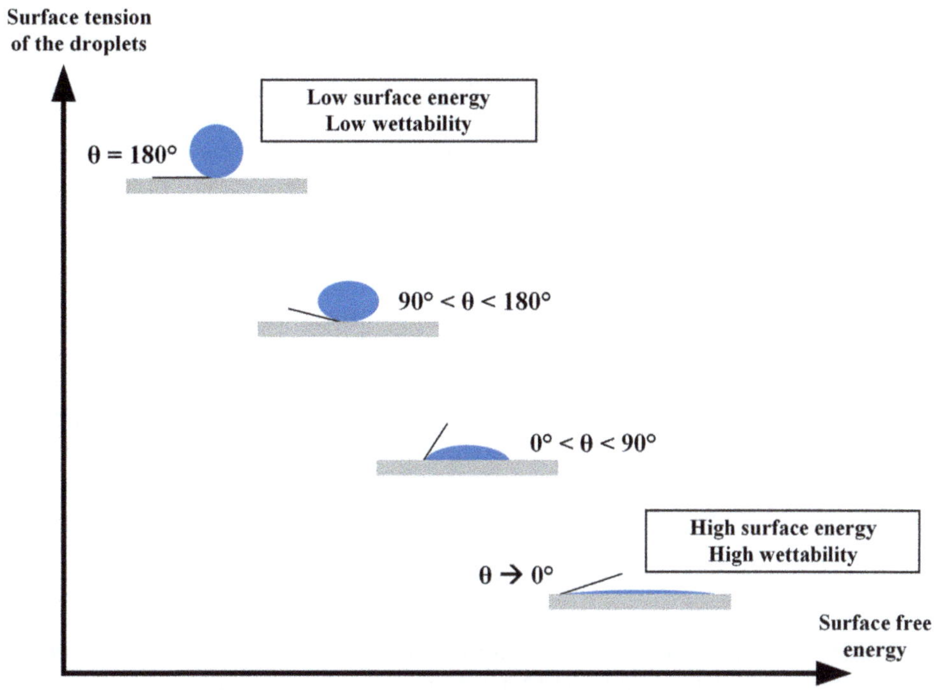

Figure 6.3. Different models of wetting of solid surfaces.

diiodomethane, nitromethane, glycerol, formamide and ethylene glycol [20–22]. This of course makes it very difficult to compare SFE calculated using different liquids and methods, especially when the object of the investigation is the modification of the surface characteristics. Most studies use the Owens–Wendt method and the two known liquids are usually water, which has polar characteristics, and diiodomethane, which has nonpolar (or dispersive) characteristics [23].

Figure 6.3 represents schematically the link between the SFE of the solid surface and its wettability.

Laser surface texturing is an effective method to change the SFE of the solid surface, often with the purpose of increasing the strength of adhesive-bonded joints, especially when polymeric substrates are involved. For example, Sun *et al* [22] estimated the SFE determining the contact angles of three test liquids on CFRP substrates using the sessile drop method. As often happens, the liquids were individuated to cover a large range of behaviors, from highly polar to almost completely dispersive. SFE was increased by about the 50%. Similarly, Wilson *et al* [24] reports the effects of laser treatment of PEEK on SFE, which increased from 44.9 to 72.5 mJ m^{-2}.

6.2.2 Investigation methodologies in industrial context

As extensively mentioned, wettability properties are typically quantified in terms of the apparent contact angle.

Contact angles are usually evaluated using a direct measurement of the static angle formed by a calibrated (i.e. with a defined volume) drop on a substrate at the three phases contact point (θ in figure 6.1). Several measurements with different liquids are required and they are usually conducted on specific goniometers. The drop is positioned on the substrate under consideration, and the angle between the tangent to the drop and the substrate is acquired using a high resolution camera and the appropriate software. In order to have reliable results, the acquisition time of angle measurements and the temperature must be controlled and stable. In fact, after some seconds the liquid could partially evaporate and the volume could not be the same for all the measurements. The acquisition software could also be used for the conversion of the contact angles to SFE values, using the desired model.

The most sophisticated goniometer could be used for laboratory and research analysis, but they are delicate and usually require fixed installation. An example of a drop analyzer and an image of a droplet as could be seen by acquisition software is reported in figure 6.4.

To have a quantitative but faster result, more suitable for industrial use, two main possibilities have been developed.

The first one is based on the traditional acquisition of contact angles, but adapted for mobile equipment, which can be used also on finished products (figure 6.5).

Another method, which is most suitable for use during production processes, involves the use of test liquids or Dyne pens. Test liquids provide the evaluation of a range of SFE depending on the producer, but of a maximum between 30 and 72 mN m^{-1} [27–29]. The test involves the application of a solution of known surface tension equivalent to the expected substrate SFE value and one examines the behavior of the liquid after 2 s. If the solution split into several drops immediately, a test with a lower tension solution must be conducted, until the liquid that is not separating is found (figure 6.6(a)).

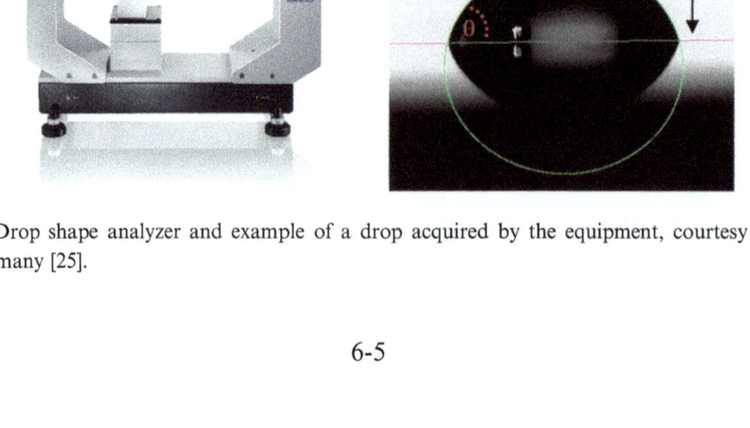

Figure 6.4. Drop shape analyzer and example of a drop acquired by the equipment, courtesy of KRÜSS GmbH, Germany [25].

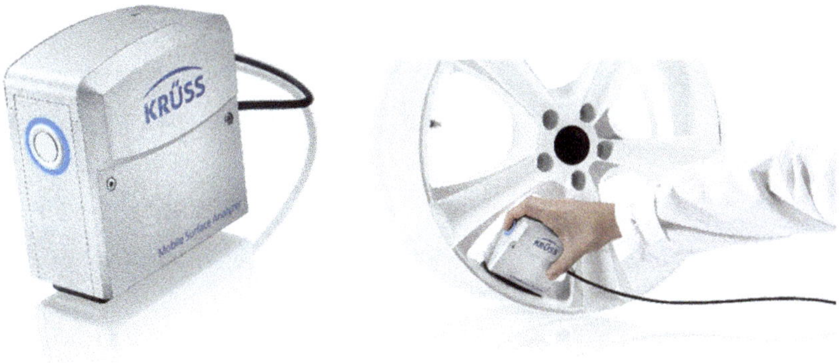

Figure 6.5. Portable drop shape analyzer. Courtesy of KRÜSS GmbH, Germany [26].

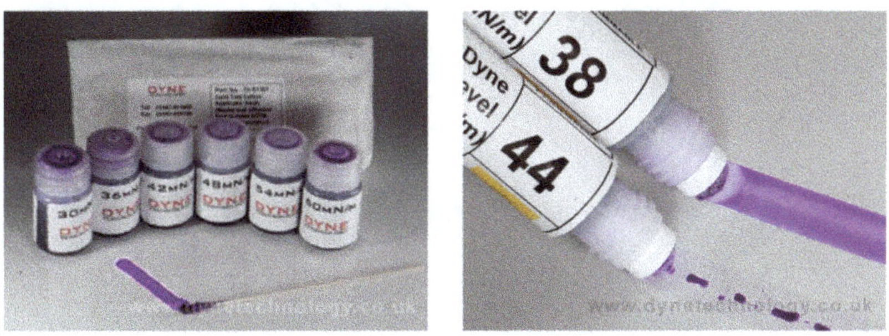

(a) (b)

Figure 6.6. Test liquids (a) and Dyne pens (b). Courtesy of Dyne testing Ltd. [27, 30].

Dyne pens have exactly the same mechanism but the different liquids are contained in a more comfortable pen, with which a line of about 5–7 cm has to be drawn (figure 6.6(b)).

6.3 Surface morphology

Surface morphology is one of the characteristics that laser texturing affects the most. It is possible to realize very complex and repetitive structures on micro- and nano-scale levels on the surface in order to reach the most suitable one for the expected application. Morphology describes surface texture, shape and distribution of the material, in a qualitative manner. Very often the topography, through a quantitative survey, is also necessary in order to compare different laser treatments using, for example, surface roughness parameters [31]. Marinescu *et al* [32] summarize and describe the most common roughness data in a figure, reproduced in figure 6.7.

These parameters can also be reported as surface quantities, and usually their designation is the same as reported in the figure but using the letter 'S', instead of 'R', for example S_a, S_p and so on.

Typical structures generated by laser processing are spikes, holes, bumps or ripples, surface structures with a periodic trend; examples of what they look like are shown in figure 6.8.

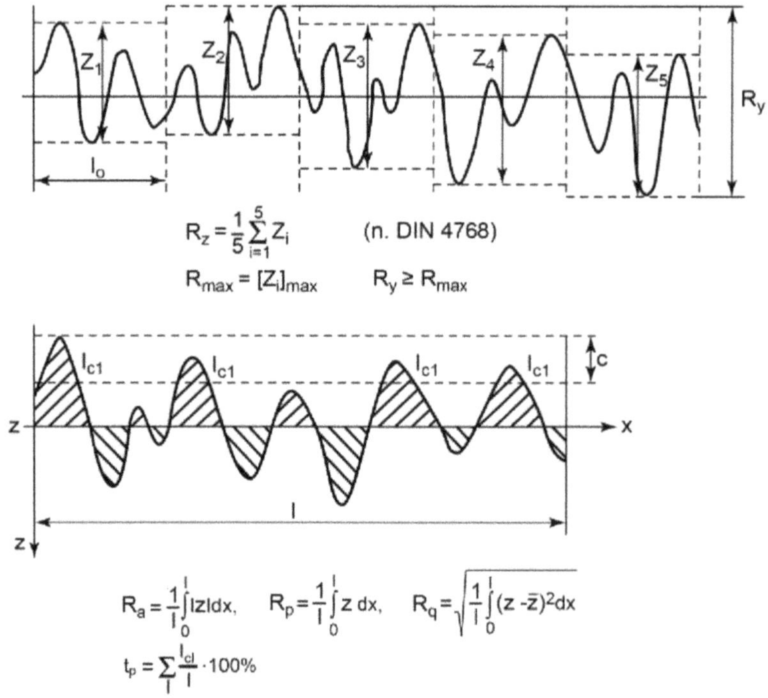

Figure 6.7. Surface roughness parameters. Reprinted from [32], copyright (2015), with permisson from Elsevier.

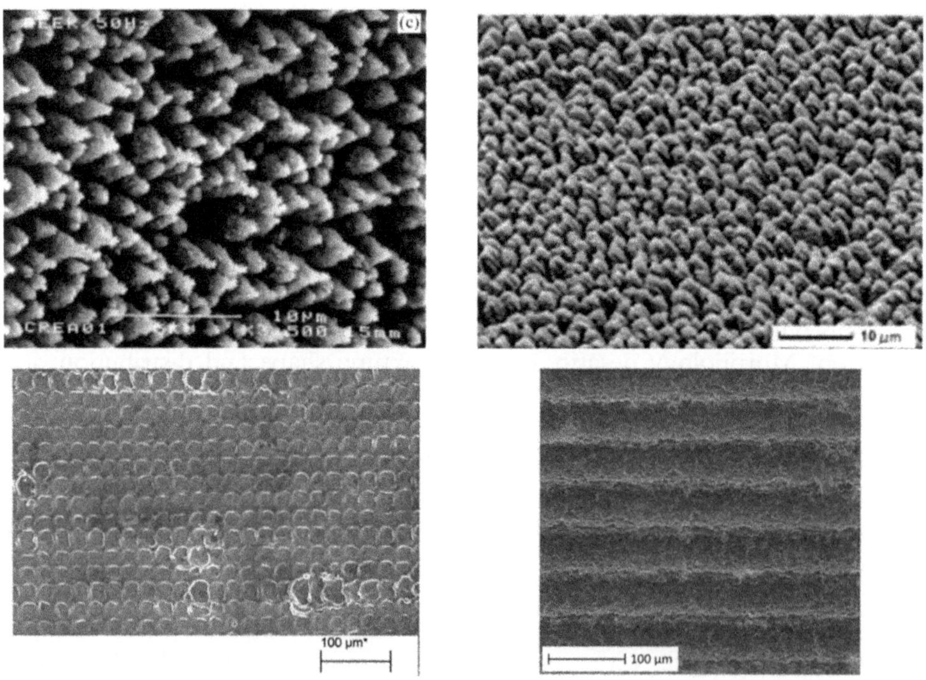

Figure 6.8. Laser textured surfaces. Reprinted from [21, 33–35], with permission from Elsevier.

Several analytical techniques are available and their choice depends on the information required (i.e. qualitative or quantitative).

Usually when the investigation is based on what a surface looks like, the nature and type of structures created, an imaging analytical instrument designed to facilitate imaging from low to high resolution could be suggested. In this case, a 3D stereo-microscope and scanning electron microscope (SEM) are the most suitable for this scope and several research works confirm this [22, 36–40].

Instead of light, SEM uses a focused beam of electrons to scan the surface and form the image of the surface morphology. In particular, secondary electrons interact with the atoms of the sample, excited by the electron beam, and they produce signals that, together with beam position, are related to the surface topography and surface composition of the sample examined. The achievable resolution is in the order of a few nanometers. The images reported in figure 6.8, for example, are all acquired by SEM.

Contact or non-contact profilometers are commonly used for both visualizing and measuring surface morphology and roughness. In the contact type, a stylus operates in contact with the surface and scans it, which could be a problem for 'soft' materials, while in the non-contact type nothing is in direct contact with the surface. Many sophisticated profilometers could be used in different modes and both in contacting or non-contacting modes. In advanced equipment, the position of the probe can be registered using a piezoelectric transducer or a laser beam.

In the contact mode, the stylus can give a vertical resolution of a few nanometers and a horizontal one ten times higher. Usually 3D images, which range from few hundreds of square microns to few millimeters square areas, are acquired.

Non-contact profilometers generally use laser technology and are able to reach a vertical resolution of the order of a nanometer. In the past, the reflective properties of some materials could be a problem for the use of this type of technology. Today, even high reflectivity surfaces can be scanned and at high speeds.

For more precise measurements on a sub-nanometer scale, the most used measuring instrument is the atomic force microscope (AFM), which reproduces the surface morphology and topography of the surface under examination. It records the interaction forces between the surface of the material itself and a probe mounted on a cantilever. It is possible to obtain a sub-nanometric resolution and to provide spatial information on the plane but also perpendicular to the surface, resulting in a 3D image of the sample surface. Examples are reported in figure 6.9.

Since it is not based on an electric current flow, AFM can be used indifferently on conductive or non-conductive surfaces and in any environment (i.e. air, vacuum or also fluid). Over the years, different modes of operation have been developed for the AFM. The three main modes of operation are classified according to the motion of the probe, namely the static mode, or contact mode, and two dynamic modes, the tapping mode and the non-contact mode.

The contact mode is based on the use of van der Waal's attractive forces between the probe and surface examined, while in the non-contact mode, the probe scans the surface vibrating at a constant distance from the surface and amplitude vibration is related to the surface morphology. In the tapping mode, the cantilever of the probe

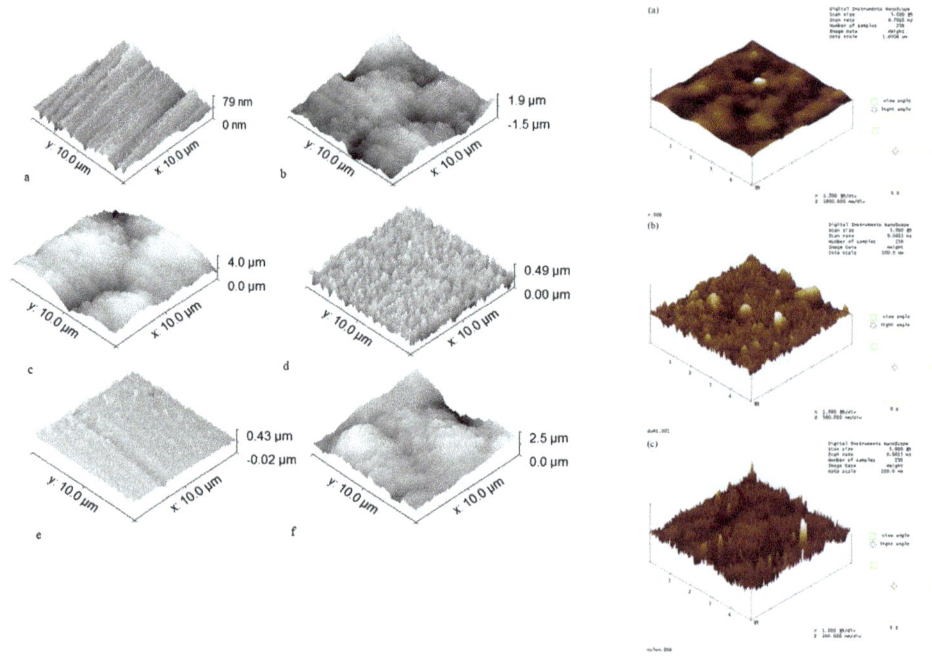

(a) (b)

Figure 6.9. AFM images of laser-treated stainless steel and a control sample (a) [41], and laser-treated PA6 films and a control sample (b) [42]. Reprinted with permission from Elsevier.

is brought to its resonance vibration or near it, such that the probe and sample are in an alternating contact/non-contact situation.

The use of AFM for the evaluation of surface texturing is reported in several studies [41–46].

6.4 Chemical composition

The interaction between a laser beam and the surface of a material could bring changes not only in wettability behavior and in morphology, but also to chemical modification, that could be time-dependent or stable.

Chemical composition can affect several surface characteristics, such as adhesion properties and corrosion resistance. The analysis of surface chemical composition can be conducted using several techniques including those involving probing species of electrons, ions and photons. The elemental composition is a crucial aspect, but also the nature of the chemical bonding of the surface atoms deserves to be investigated, which is particularly useful to create a stable link to other materials.

The most commonly used analysis uses X-ray photoelectron spectroscopy (XPS) or infrared (IR) spectroscopy (FTIR).

XPS is a surface analysis technique able to provide elemental composition, chemical state and chemical bonds. The sample surface under investigation is placed in a vacuum chamber and bombarded by X-rays. The energy of the photoelectrons (in the

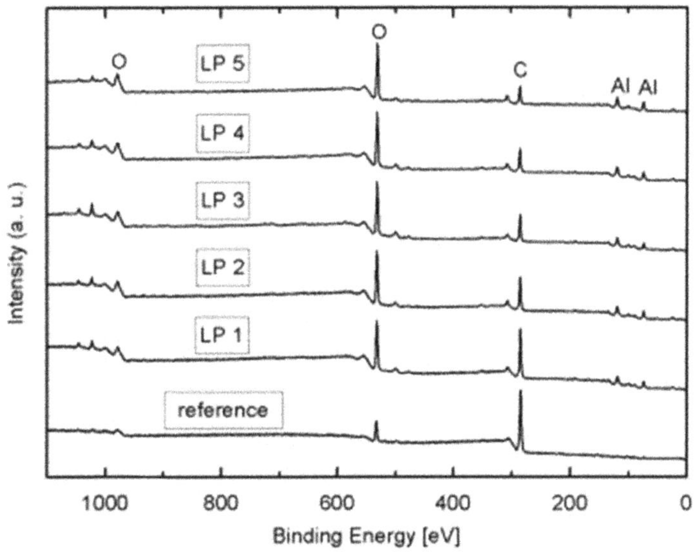

Figure 6.10. XPS analysis for chemical composition evaluation of untreated and laser-treated aluminum AW6016 surfaces. Reprinted from [47], copyright (2010), with permisson from Elsevier.

range of 50–2000 eV) emitted by the surface (the topmost layer of 1–10 nm thickness) and, at the same time, the kinetic energy of the same electrons are measured. In other words, the result is a photoelectron spectrum acquired by counting the emitted electrons in a specific range of electron kinetic energies. The spectrum presents several peaks representing the percentage of atoms emitting electrons of a specific energy. Putting together the information of electron intensities and energies, elements' identification and their quantification is possible. An example of XPS conducted on untreated and laser-treated samples is reported in figure 6.10.

Different X-rays, both monochromatic and non-monochromatic beams, can be used for the analysis and the nature of the photoelectron is related to this aspect. The most used is the mono-energetic Kα X-ray radiation from aluminum (1486.6 eV) or magnesium (1253.6 eV). The energy of the photoelectron is usually detected by an electron energy analyzer.

As mentioned above, also chemical bonding can be measured from the emission spectra. An evaluation can be obtained observing the so-called 'chemical shifts' of the energy position of the electrons in the XPS spectra. Typical examples are carbon and oxygen bonding analysis: XPS detects carbon and/or oxygen elements and focuses on the elements' spectra (the range immediately close to the specific energy); from the shifts it is possible to understand what kind of bonds are present and in what percentage.

Figure 6.11 compares XPS survey spectra of untreated and laser-treated carbon fiber reinforced with epoxy resin [22]. The spectra show a reduction of C–O bond content from the control to the laser-treated sample, which the researchers relate to a decrease of adhesion between the epoxy resin and the carbon fiber.

Many researchers rely on the analysis of chemical surface changes by another method, Fourier transform infrared spectroscopy (FTIR analysis or spectroscopy).

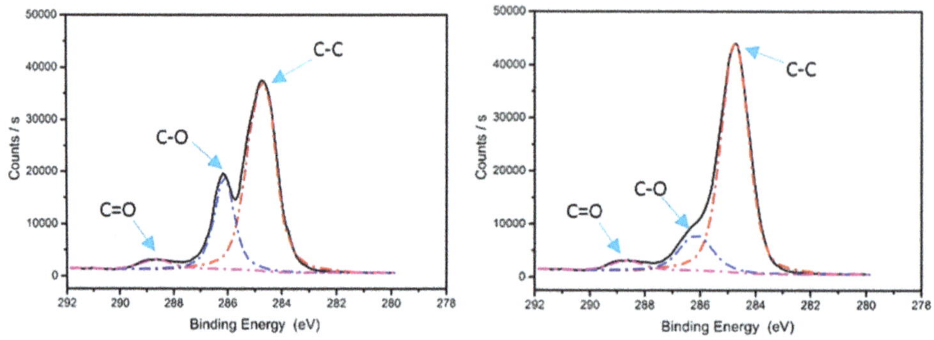

Figure 6.11. XPS survey spectra for the evaluation of carbon bonds of untreated and laser-treated CFRP surfaces. Reprinted from [22], copyright (2018), with permisson from Elsevier.

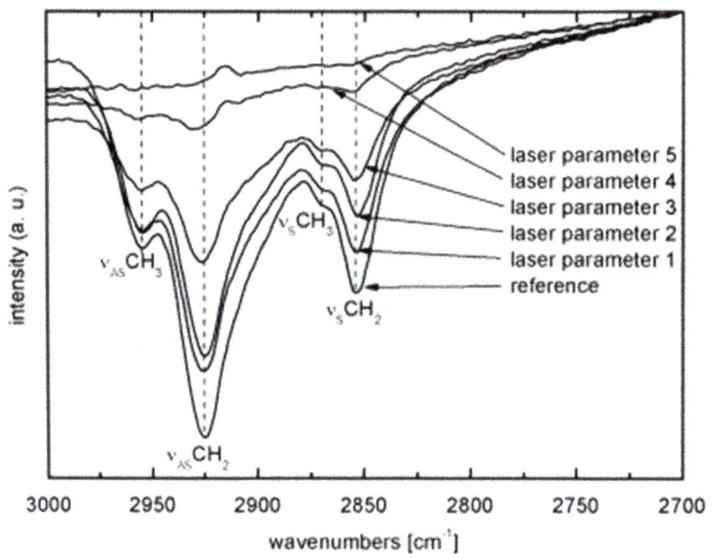

Figure 6.12. FTIR spectrum of differently treated aluminum. Reprinted from [47], copyright (2010), with permisson from Elsevier.

The main purpose of this technique is to identify organic or inorganic materials, and very often it is used on polymeric samples. The basic principles of FTIR is the use of infrared light to scan test samples. Some radiation is absorbed and some passes through the material. The molecules of the sample convert the absorbed part of the radiation into vibrational energy, the signal is transmitted to a detector and the results are visualized as a spectrum, representing a sort of 'fingerprint' of the material. It allows the chemical identification of an element or species. For example, in figure 6.12, FTIR analysis conducted by Rechner *et al* [47] on aluminum substrates demonstrates the cleaning properties of laser surface treatment. In fact, while the reference is contaminated with organic components (absorption of the signal in hydrocarbon compounds), with proper laser parameters CH bands are no longer detected.

The most difficult aspect is the interpretation of the data, even if the information that could be obtained with this analysis is quite considerable. In fact, the total spectrum is formed by a series of values of absorbed energy. The spectrum generated is rather discrete, constituted by bands of energy. The trend of the peaks of energy is dependent on a series of factors, chemical and related to the incident energy introduced. Hence, the correspondence between the energy band and the chemical bond is often not unequivocal. The interpretation of the spectrum is conducted by analysts with specific experience.

6.5 Durability behavior

In order to be able to fully exploit the characteristics imparted to the surface through laser treatment, it is necessary to evaluate them not only in the condition of just-performed treatment, but also in their stability over time.

Environmental action can significantly affect the performance of the surface. Tests that assess how surface characteristics change under exposition to different environments usually focus on the variation of joint strength [34, 39, 47, 48], generally keeping the samples in controlled conditions of temperature and humidity for some days.

Regarding surface treatments that increase the adhesion characteristics of surfaces, they are typically evaluated based on the variation of the performance of bonded joints. Ebnesajjad and Landrock summarize the different kinds of tests to which adhesive-bonded joints could be subjected. As they specify, the evaluation is usually done not only on the mechanical strength properties, but also on the bonding technique and the efficiency of surface treatments [49].

Analysis on durability is usually carried out by experimental comparison, but every standard states that none of the laboratory tests are appropriate for a quantitative estimate of real durability. The usual procedure dictates that samples undergo short-term (accelerated) aging in environments properly recreated in laboratory, and a comparison is made with the characteristics of non-aged samples.

Even if the list of possible tests is quite extensive, because it has the purpose of covering different applications, the tests actually reported in research works are relatively few.

Very often the reference is to the EN ISO 9142 standard [50], in which critical environmental conditions, performed as accelerated aging cycles, are reported. In particular, a simulation of atmospheric conditions was performed controlling temperature humidity, and sometimes immersion in water periods, that are detected as the most critical condition.

Examples of the studies conducted, even on laser surface texturing, have highlighted how an adequate surface preparation actually makes the difference even on the long-term performance of bonded joints [34]. Figure 6.13 compares the characterization of tensile shear strength of DP 500 and AISI 304 adhesive-bonded joints before and after environmental aging [48].

One of the most significant tests to compare different surface preparations is the wedge test [51]. The test procedure requires that a wedge is forced at a constant speed into an adhesive-bonded specimen in correspondence of the bond line (figure 6.14).

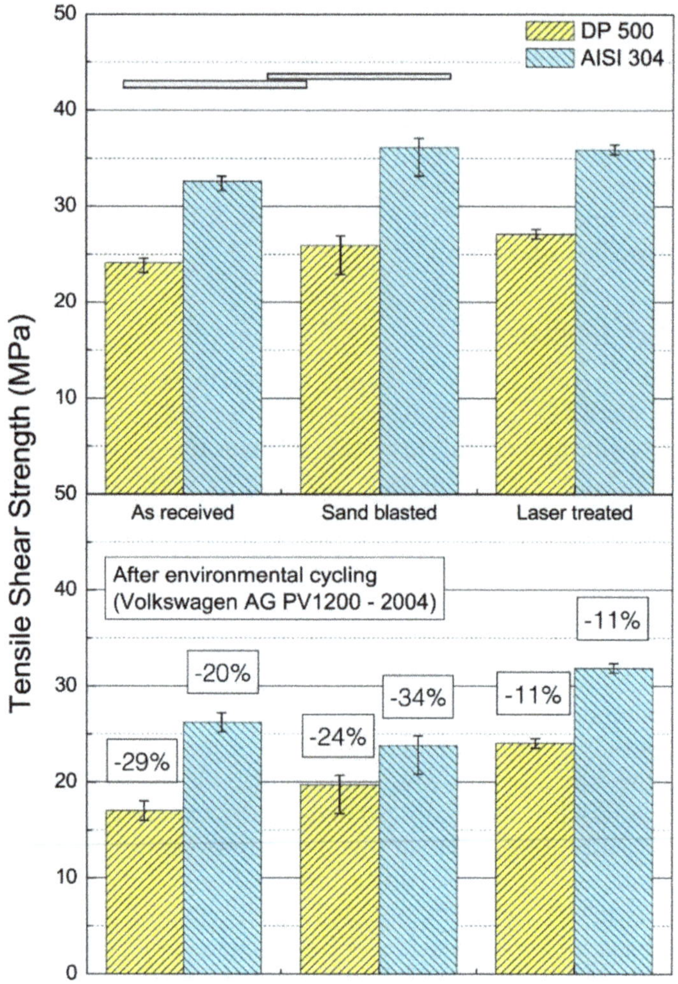

Figure 6.13. Comparison of differently treated DP 500 and AISI 304 epoxy joints before and after environmental cycling. Reprinted from [48], copyright (2015), with permisson from Elsevier.

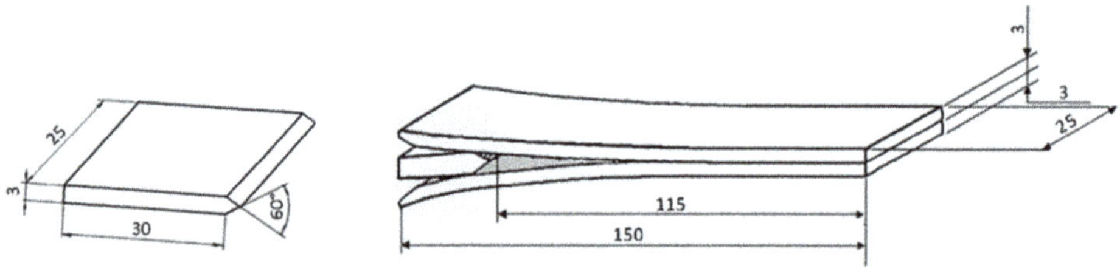

Figure 6.14. Wedge and specimen configuration according to ASTM D3762. Reprinted from [52], copyright (2018), with permisson from Elsevier.

A crack is generated and measured; additionally, and the presence of the wedge creates a stress in the area of the crack tip. The specimen could now be exposed to specific environments (high temperature, cycle of changing temperature and humidity and so on) always keeping the wedge inserted. In this way the combination of mechanical and environmental stresses causes the growth of the crack, the trend of which is periodically evaluated during the exposition. At the end of the test the substrates have to be separated and the fracture surfaces observed to understand the failure mechanism. The test is considered qualitative and comparative, but it is a remarkable method to determine the soundness of adherend surface preparation and adhesive system (including substrates, adhesive and interface) durability.

6.6 Conclusions

The changes introduced on the surface through the interaction of a laser beam with different kinds of materials need to be investigated with specific and accurate methods that can highlight aspects that are useful for the industrial application of the modifications themselves.

In this chapter, the main and most commonly used methods of superficial investigation have been illustrated, with a mention also of the evaluation of the stability characteristics over time of the same laser treatments, which take on particular importance when applied to adhesive-bonded joints.

References

[1] Grundke K, Pöschel K, Synytska A, Frenzel R, Drechsler A and Nitschke M *et al* 2015 Experimental studies of contact angle hysteresis phenomena on polymer surfaces—Toward the understanding and control of wettability for different applications *Adv. Colloid Interface Sci.* **222** 350–76

[2] Kubiak K J, Wilson M C T, Mathia T G and Carval P 2011 Wettability versus roughness of engineering surfaces *Wear* **271** 523–8

[3] Łęcka K M, Gąsiorek J, Mazur-Nowacka A, Szczygieł B and Antończak A J 2019 Adhesion and corrosion resistance of laser-oxidized titanium in potential biomedical application *Surf. Coatings Technol.* **366** 179–89

[4] Boinovich L and Emelyanenko A 2011 Wetting and surface forces *Adv. Colloid Interface Sci.* **165** 60–9

[5] Riveiro A, Maçon A L B, del Val J, Comesaña R and Pou J 2018 Laser surface texturing of polymers for biomedical applications *Front. Phys.* **6** 16

[6] Nanoscience Instruments, USA website https://nanoscience.com/techniques/tensiometry/surface-free-energy/ accessed 4 Sept 2019

[7] Packham D E 2005 *Handbook of Adhesion* vol 1 2nd edn (Hoboken, NJ: Wiley) pp 1–638

[8] Barati Darband G, Aliofkhazraei M, Khorsand S, Sokhanvar S and Kaboli A 2018 Science and engineering of superhydrophobic surfaces: review of corrosion resistance, chemical and mechanical stability *Arab. J. Chem.* **13** 1763–802

[9] Wenzel R N 1936 Resistance of solid surfaces to wetting by water *Ind. Eng. Chem.* **28** 988–94

[10] Cassie B D and Baxter S 1944 Wettability of porous surfaces *Trans. Faraday Soc.* **40** 546–51

[11] Kubiak K J, Wilson M C T, Mathia T G and Carras S 2011 Dynamics of contact line motion during the wetting of rough surfaces and correlation with topographical surface parameters *Scanning* **33** 370–7
[12] Belaud V, Valette S, Stremsdoerfer G, Bigerelle M and Benayoun S 2015 Wettability versus roughness: multi-scales approach *Tribol. Int.* **82** 343–9
[13] Webb H K, Crawford R J and Ivanova E P 2014 Wettability of natural superhydrophobic surfaces *Adv. Colloid Interface Sci.* **210** 58–64
[14] Rudawska A and Jacniacka E 2009 Analysis for determining surface free energy uncertainty by the Owen–Wendt method *Int. J. Adhes. Adhes.* **29** 451–7
[15] Cao G and Wang Y 2011 Physical chemistry of solid surfaces *Nanostructures and Nanomaterials Synthesis, Properties, and Applications* 2nd edn ed M Reed (Singapore: World Scientific) 16–60
[16] Packham D E 2003 Surface energy, surface topography and adhesion *Int. J. Adhes. Adhes.* **23** 437–48
[17] Fowkes F M 1963 Additivity of intermolecular forces at interfaces. I. Determination of the contribution to surface and interfacial tensions of dispersion forces in various liquids *J. Phys. Chem.* **67** 2538–41
[18] Owens D K and Wendt R C Estimation of the surface free energy of polymers *J. Appl. Polym. Sci.* **13** 1741–7
[19] Wu S 1971 Calculation of interfacial tension in polymer systems *J. Polym. Sci. Part C. Polym. Symp.* **34** 19–30
[20] Encinas N, Díaz-Benito B, Abenojar J and Martínez M A 2010 Extreme durability of wettability changes on polyolefin surfaces by atmospheric pressure plasma torch *Surf. Coatings Technol.* **205** 396–402
[21] Laurens P, Sadras B, Decobert F and Amouroux J 1998 Enhancement of the adhesive bonding properties of PEEK by excimer laser treatment *Int. J. Adhes. Adhes.* **18** 19–27
[22] Sun C, Min J, Lin J, Wan H, Yang S and Wang S 2018 The effect of laser ablation treatment on the chemistry, morphology and bonding strength of CFRP joints *Int. J. Adhes. Adhes.* **84** 325–34
[23] Izdebska J 2016 Printing on polymers: theory and practice *Printing on Polymers: Fundamentals and Applications* ed J Izdebska and S Thomas (Boston, MA: William Andrew Publishing) 1–20
[24] Wilson A, Jones I, Salamat-Zadeh F and Watts J F 2015 Laser surface modification of poly (etheretherketone) to enhance surface free energy, wettability and adhesion *Int. J. Adhes. Adhes.* **62** 69–77
[25] KRÜSS GmbH, Germany website https://kruss-scientific.com/products/drop-shape/dsa30/drop-shape-analyzer-dsa30/ accessed 4 Sept 2019
[26] KRÜSS GmbH, Germany website https://kruss-scientific.com/products/drop-shape/mobile-surface-analyzer-msa/ accessed 17 Sept 2019
[27] Dyne Testing Ltd, Staffordshire, UK website http://dynetechnology.co.uk/products/dyne-test-fluids-fluid001/ accessed 4 Sept 2019
[28] Ferrarini & Benelli Srl, Italy website https://it.ferben.com/Prodotti/Altri-Prodotti/Inks.kl accessed 4 Sept 2019
[29] Tantec A/S, Denmark website https://tantec.com/dyne-test-measuring-surface-energy.html accessed 4 Sept 2019
[30] Dyne Testing Ltd, Staffordshire, UK website http://dynetesting.com/surface-energy-measurement/dyne-pens/ accessed 17 Sept 2019

[31] Mattox D M 1998 Substrate ('real') surfaces and surface modification *Handbook of Physical Vapor Deposition (PVD)* (Boston, MA: William Andrew Publishing) pp 56–126

[32] Marinescu I D, Rowe B, Ling Y and Wobker H G 2015 Abrasive processes *Handbook of Ceramics Grinding and Polishing* 2nd edn ed I D Marinescu, T K Doi and E Uhlmann (Boston, MA: William Andrew Publishing) 67–132

[33] Li B-j, Li H, Huang L-j, Ren N-f and Kong X 2016 Femtosecond pulsed laser textured titanium surfaces with stable superhydrophilicity and superhydrophobicity *Appl. Surf. Sci.* **389** 585–93

[34] Wu Y, Lin J, Carlson B E, Lu P, Balogh M P and Irish N P *et al* 2016 Effect of laser ablation surface treatment on performance of adhesive-bonded aluminum alloys *Surf. Coatings Technol.* **304** 340–7

[35] Moroni F, Romoli L and Khan M M A 2018 Design of laser-textured surfaces to enhance the strength of adhesively bonded joints *Int. J. Adhes. Adhes.* **85** 208–18

[36] Palmieri F L, Ledesma R I, Dennie J G, Kramer T J, Lin Y and Hopkins J W *et al* 2019 Optimized surface treatment of aerospace composites using a picosecond laser *Compos. Part B Eng.* **175** 107155

[37] Grabowski A, Sozańska M, Adamiak M, Kępińska M and Florian T 2018 Laser surface texturing of Ti6Al4V alloy, stainless steel and aluminium silicon alloy *Appl. Surf. Sci.* **461** 117–23

[38] Li J, Zhou Y, Fan F, Du F and Yu H 2019 Controlling surface wettability and adhesive properties by laser marking approach *Opt. Laser Technol.* **115** 160–5

[39] Rotella G, Orazi L, Alfano M, Candamano S and Gnilitskyi I 2017 Innovative high-speed femtosecond laser nano-patterning for improved adhesive bonding of Ti6Al4V titanium alloy *CIRP J. Manuf. Sci. Technol.* **18** 101–6

[40] Rauh B, Kreling S, Kolb M, Geistbeck M, Boujenfa S and Suess M *et al* 2018 UV-laser cleaning and surface characterization of an aerospace carbon fibre reinforced polymer *Int. J. Adhes. Adhes.* **82** 50–9

[41] Rajab F H, Liauw C M, Benson P S, Li L and Whitehead K A 2018 Picosecond laser treatment production of hierarchical structured stainless steel to reduce bacterial fouling *Food Bioprod. Process.* **109** 29–40

[42] Gao Z, Peng S, Sun J, Yao L and Qiu Y 2009 Influence of processing parameters on atmospheric pressure plasma etching of polyamide 6 films *Appl. Surf. Sci.* **255** 7683–8

[43] Brack N and Rider A N 2014 The influence of mechanical and chemical treatments on the environmental resistance of epoxy adhesive bonds to titanium *Int. J. Adhes. Adhes.* **48** 20–7

[44] Gao Z, Peng S, Sun J, Yao L and Qiu Y 2010 The influence of moisture on atmospheric pressure plasma etching of PA6 films *Curr. Appl. Phys.* **10** 230–4

[45] Rebollar E, Castillejo M and Ezquerra T A 2015 Laser induced periodic surface structures on polymer films: from fundamentals to applications *Eur. Polym. J.* **73** 162–74

[46] Bizi-bandoki P, Benayoun S, Valette S, Beaugiraud B and Audouard E 2011 Modifications of roughness and wettability properties of metals induced by femtosecond laser treatment *Appl. Surf. Sci.* **257** 5213–8

[47] Rechner R, Jansen I and Beyer E 2010 Influence on the strength and aging resistance of aluminium joints by laser pre-treatment and surface modification *Int. J. Adhes. Adhes.* **30** 595–601

[48] Rotella G, Alfano M, Schiefer T and Jansen I 2015 Enhancement of static strength and long term durability of steel/epoxy joints through a fiber laser surface pre-treatment. *Int. J. Adhes. Adhes.* **63** 87–95

[49] Sina Ebnesajjad A H L 2009 Durability of adhesive bonds *Adhesives Technology Handbook* 2nd edn (Boston, MA: William Andrew Publishing) p 231–72

[50] CEN 2004 *Adhesives. Guide to selection of standard laboratory ageing conditions for testing bonded joints* (European Committee for Standardization, CEN) BS EN ISO 9142:2003

[51] ASTM International 2006 *Adhesive-Bonded Surface Durability of Aluminum (Wedge Test) ASTM D3762* vol 03 (West Conshohocken, PA: ASTM International) pp 3–7

[52] Mandolfino C, Lertora E, Gambaro C and Pizzorni M 2018 Durability of polyamide bonded joints: influence of surface pre-treatment *Int. J. Adhes. Adhes.* **86** 123–30

Chapter 7

Modelling of laser micro-processing techniques

Israt Rumana Kabir and Sumsun Naher

List of symbols and abbreviations

Symbols	Specification	Unit
P	Laser power	W
C_P	Specific heat	J kg^{-1} K^{-1}
K	Thermal conductivity	W m^{-1} K^{-1}
T	Temperature	K or °C
T_0	Initial temperature	K or °C
Q	Internal heat generation	W m^{-3}
Q_{laser}	Laser heat intensity	W m^{-2}
h	Convection film coefficient	W m^{-2} K^{-1}
A	Absorptivity	—
r_b	Laser beam radius	m
t_R	Residence time	s
t_C	Time of cooling	s
ρ	Density	kg m^{-3}
ε	Emissivity	—
σ	Boltzmann's constant, 5.67×10^{-8}	W m^{-2} K^{-4}
π	Pi constant, 3.1416	—
$\varepsilon^{\text{total}}$	Total strain	—
ε^{el}	Elastic strain	—
ε^{pl}	Plastic strain	—
ε^{th}	Thermal strain	—
q_X	Heat flux rate along X-axis	W m^{-2}
r_{beam}	Radius of laser beam	m
α	Coefficient of thermal expansion	K^{-1}

(*Continued*)

(*Continued*)

Symbols	Specification	Unit
$V_{y,i}$	Volume fraction of *y*-phase at *i*th step	%
V_{aus}	Volume fraction of austenite	%
V_{marten}	Volume fraction of martensite	%
M_s	Martensite-start transformation temperature	K
	Abbreviations	
LSG	Laser surface glazing	
LW	Laser welding	
GTAW	Gas tungsten arc welding	
LC	Laser cladding	
SLM	Selective laser melting	
LSM	Laser surface melting	
FEM	Finite element method	
FDM	Finite difference method	
FVM	Finite volume method	
TMM	Transitional mapped meshing	
APDL	ANSYS parametric design language	
TMT	Thermomechanical treatment	
HAZ	Heat-affected zone	
CW	Continuous wave	
T–T–T	Time–temperature–transformation	
FEA	Finite element analysis	
TEM	Transverse electromagnetic	
SF	Surface force	
BF	Body force	
HPDDL	High power direct diode laser	
Nd:YAG	Neodymium-doped yttrium aluminium garnet	

7.1 Introduction

This chapter has been developed to discuss the progress in numerical thermal thermomechanical modelling of laser melting processes including laser surface glazing (LSG), laser welding (LW), laser cladding (LC) and selective laser melting (SLM). Hardness modelling of laser modification techniques has briefly been discussed. After that, the salient controlling factors in the development of numerical models for the laser melting processes have also been outlined from the literature survey.

7.1.1 Thermal modelling of laser surface glazing and similar processes

Over the last decade, there have been numerous thermal modelling works of laser surface modification processes carried out. The major purpose of thermal modelling is to study the underpinned physical phenomena, predict temperature distribution [1] and thermal effects [2, 3] on the surface of the treated part and subsequently to optimise the processes [4]. Through thermal modelling, extensive numbers of trials can be performed with reduced cost and time unlike experimental procedures. The physics of laser surface modification techniques, for their very fast nature, deviates from the equilibria. Therefore, the heat transfer phenomena of such processes are complex to describe with simple analytical modelling. Furthermore, along with a wide range of process parameters, laser absorption and material properties, all play key roles in regulating heat transfer phenomena in laser surface modification techniques. In this respect, the numerical thermal modelling technique has overcome the problem for the research and development of laser surface modification processes. This section of the chapter will review some recent numerical thermal modelling works of LSG, LW, LC and SLM to explore the parameters, materials and method used for those models and the outcome of those modelling works.

7.1.1.1 Laser surface glazing

Since LSG has been introduced, very few modelling works of LSG for metallic materials have been carried out. A single numerical model of LSG for metallic and ceramic thermal barrier coating was developed by Mahank in 2004 [5]. The 2D thermal model of LSG was carried out using the FDM method and predicted temperature distribution and the dimension of laser glazed tracks. Temperature-dependent material properties and variable absorptivity of metallic and ceramic materials were employed in this model. Later in 2011 [6], a thermal model of LSG of H13 tool steel to predict temperature distribution and the dimension of the melt pool was proposed using an explicit solution. The model used the constant properties of steel and predicted a nominal temperature distribution of the treated surface [7]. The study showed the isothermal temperature profiles on the XY plane at five different thicknesses along the Z-direction from this model. The nominal temperature distribution was used in plotting the heating and cooling cycle at different depths from the surface. Besides, there are few analytical models considering the modes of heat transfers conducted to predict melt pool geometry in laser glazing of ceramic materials [8–10]. All these models have a single primary goal which is predicting the temperature distribution, because the temperature is the decisive factor to the outcome of all laser melting processes. The temperature profile and temperature history obtained from the simulation are utilised to further investigate the dimension of the laser modified zone, possible phase transformation, hardness properties, thermal stress, distortion, cracks and so on. Therefore, it is very important to develop a reliable model that can predict temperature distribution and other outcomes.

7.1.1.2 Laser Welding

Laser welding is a non-contact welding process to join metal pieces by scanning a high power laser beam over the weld zone. The substantial number of modelling works in laser welding [11–13] manifests intense research and development programmes in this field. Temperature history, dimension of the fusion zone, HAZ [1, 14, 15], weld shape geometry [4, 16] and the optimisation of laser welding process [13] have been predicted through FEM thermal modelling. Different software such as ABAQUS [34], SYSWELD [12] and ANSYS [17] have been utilised for developing the model and calculating the output parameters. In the LW process, a high range of laser power 2.5–8 kW has been observed to process steels, whereas non-ferrous alloys, for example, Ti6Al4V, Zircaloy 4 (Zr–1.4Sn–0.1Cr–0.2Fe), were operated with lower laser power. An exception in the non-ferrous alloy is the aluminium alloy, which was welded at 3 kW laser power. Laser scanning speed was also observed to vary from 4 to 40 mm s^{-1} and laser spot size ranges from 0.2 to 1 mm depending on materials and processes. Different laser types were used for the high power requirement of LW processes, among them the CO_2 laser, Yb-YAG (ytterbium–-yttrium aluminium garnet), fibre and Nd:YAG lasers exert high power in a descending manner. Moreover, some hybrid LW combined with arc current [18], MIG [14] or GTAW [17] processes were also modelled for various metals and geometries. The comparison between pure LW and hybrid welding was carried out through modelling [11]. Table 7.1 shows the laser parameters used in laser welding processes for different metals and alloys. From the thermal modelling of LW, the temperature distribution, depth of weld pool and HAZ, and cooling rates have been investigated. The effect of laser power, scanning speed on temperature distribution, size of fusion zone and cooling rates also have been analysed.

7.1.1.3 Laser cladding

Laser cladding is a surface modification technique where external material in powder or wire form is deposited on a substrate utilising a scanning laser beam, the additional material is called clad [20]. The mixture of clad and substrate at the interface is called the dilution zone. The temperature distribution in clad and substrate, dimension of clad and cooling rates were predicted from different FEM thermal models of LC [2, 3, 21, 22]. The depth of dilution can also be predicted and controlled by optimising parameters through modelling [23]. In thermal modelling of LC, the range of processing power was observed to vary from 0.6 to 4 kW depending upon materials and geometries. Unlike LW, beam width or spot size differs from 1 mm [2] up to 12 mm of both spherical [23] and rectangular shape [3, 21] of the laser beam. Table 7.2 summarises the technical information and key findings from recent thermal modelling works of LC. The materials used for clad and substrate, laser type, laser parameters, numerical methods and software used have been listed. The effects of laser power, scanning speed in peak temperature in the clad, cooling rates and width of melt pool were analysed in those models. The relationship between the melt pool width and dilution zone was also analysed through modelling. Comparing table 7.1 and table 7.2 it is understood that a higher level of laser power and faster scanning speed were applied in LW than LC. However, these values of parameters mainly depend on the type of material being treated. Both spherical and rectangular laser beams were used in the LC processes. A larger beam diameter was used in LC whereas a comparatively small diameter was applied in the LW process. The size of beam diameter determines the total heat flux applied and the scanning speed is adjusted accordingly.

Table 7.1. Thermal models of the LW process for various materials and laser parameters including key findings.

	Laser parameters						
Materials/laser type	Power (kW)	Beam width (mm)	Scanning speed/RT[1] (mm s^{-1}, s)	Method/software	Findings	Year	Ref.
6065 Al alloy, Nd:YAG, CW	3	0.45	10^{-3}	FEM, ABAQUS	Fluctuations of the absorbed laser beam power generate cooling speed deviations leading to the distortion	2013	[13]
Zircaloy 4, Nd:YAG, pulse	0.3–0.9	1	4	FEM, SYSWELD	With increasing power, the size of the weld pool increases, 9% accuracy is reported of the numerical model	2012	[12]
Ti6Al4V	0.5–1.5	—	5.2	FEM, ANSYS 14.5	The peak temperature increases with laser power nonlinearly, bead geometry also changes with laser beam power	2014	[15]
Ti6Al4V, CO_2, CW	1.7	—	13.34,40	FEM, SYSWELD	With increasing laser speed, the size of the weld bead decreases	2015	[4]
Ti6Al4V, Nd:YAG, pulse	0.24	0.7	3–9	FVM, SIMPLE algorithm	With decreasing laser welding speed, the peak temperature increases and the penetration depth increases. The numerical error was found maximum 17%	2014	[1]
Alloy steel, CO_2, CW	8	0.5	25	FEM, SYSWELD	At 8 kW and 25 mm s^{-1}, the fusion zone and HAZ cool down at 100 K s^{-1} average cooling rate	2014	[19]

1. RT = residence time; 2. CW = continuous wave.

Table 7.2. Thermal models of the LC process with different materials and process parameters including key findings.

Clad/substrate	Laser type	Laser parameters			Method, software	Findings	Year	Ref.
		Power (kW)	Beam width (mm)	Scanning speed (mm s^{-1})				
H13 TS/mild steel	HPDDL (975 nm)	3.8	12 × 3	3–7	FEM, ANSYS	Peak temperature reaches 1850 °C at 5 mm s^{-1} speed, 370 °C higher than the melting point of H13. With increasing scanning speed, the cooling rate increases, and size of melt pool decreases	2014	[3]
Stellite (Co–Cr alloy)/mild steel	Nd:YAG, CW	0.6–4	3–4.5	5–20	FEM, COMSOL	Melt pool width is smaller than the critical value which is about 90% of the laser beam width, the dilution level is small	2011	[23]
Ti6Al4V/Ti6Al4V	Nd:YAG	0.4–0.6	0.8–1.8	8.33–11.66	FEM, ANSYS	With increasing laser power, the depth of melt pool increases, error in predicted results is claimed maximum 3.16% than the experimental	2013	[20]
H13 TS/AISI 4140 steel	HPDDL (808 nm)	1.4–1.8	12 × 1	6–9	FEM, ANSYS 11	The average peak temperature and cooling rate are predicted 1729 °C and 1300 °C s^{-1}, respectively, in clad of H13, for power 1.6 kW and 7 mm s^{-1} speed	2011	[21]

7.1.1.4 *Selective laser melting*

Selective laser melting is a process of rapid prototyping which creates mechanical parts using a layer-by-layer scanning laser beam over the powder beds of materials [24]. A significant number of numerical thermal modelling of SLM processes are also published in this field. Table 7.3 presents an information summary of some of the numerical thermal models of SLM processes. This table includes types of materials and lasers, process parameters, and key findings from those models. SLM processes use a distinctly smaller beam width ranging from 0.01 to 0.3 mm. Compared with other laser melting processes, lower laser power from 110 W to 320 W and higher scanning speeds varying from 200 to 1939 mm s^{-1} were reported to operate SLM for Ti6Al4V, aluminium (Al) and a copper–ceramic composite [25–27]. Thermal modelling of SLM can predict the temperature distribution, dimension of the melt pool and optimise the process parameters [25]. The relative density and pore size can also be calculated through thermal modelling of SLM, which requires coupling of fluid flow dynamics with the heat transfer mechanism [28].

From reviewing those modelling works of laser melting processes the possible outcome and state-of-the-art of numerical laser modelling have been realised. The effects of some parameters including types of material and laser heat source to develop successful numerical thermal models of laser melting processes have also been understood, which will be described in subsequent sections. It is obvious that very high peak power and cooling rates are achieved in those processes, which eventually lead to induce thermal residual stress in the treated parts. The temperature distribution data from the thermal model can be further used to predict the residual stress through thermomechanical models of laser melting processes.

7.1.2 Residual stress of laser melting processes

The localised melting of the surface and remaining cold bulk material result in a steep thermal gradient and rapid cooling rates (10^5–10^8 K s^{-1}) [29, 30] in the laser melting processes. This large temperature gradient causes non-uniform thermal expansion and contraction between the surface and the bulk, which creates irregular distribution of thermal load inside the treated parts. Moreover, the rapid cooling rates encourage diffusion-less phase transformations, which incur micro strain in the treated zone due to the difference in volume of the phases. These phenomena eventually lead to the formation of residual stresses in the laser-treated region, interface and the bulk. Dai *et al* [28] reported that a 3.9×10^5 K m^{-1} temperature gradient was observed during the SLM process of Cu–WC composite, which was indicative of the presence of a significantly high stress field leading to crack formation depending on the type of residual stress. Residual stress in the laser-treated surface can be tensile or compressive in nature. It diverges in type and magnitude for different laser melting processes, materials and parameters. For instance, compressive residual stress was found in a laser cladding process operated by a fibre coupled diode laser both in continuous wave and pulse modes. In this process, powder H13 tool steel was cladded on a tempered H13 tool steel substrate. It was reported that the magnitude of compressive residual stress in the clad was

Table 7.3. Thermal models of SLM or LSM for various materials and process parameters including key findings.

		Laser parameters						
Material	Laser type	Power (kW)	Beam width (mm)	Scanning speed (mm s^{-1})	Method/software	Findings	Year	Ref.
Ti6Al4V powder	Yb fibre, CW	0.11	0.034	200	FEM, ANSYS 10	0.11 kW laser power and 200 mm s^{-1} is selected by simulation of SLM process, which gives maximum depth, 45 μm of molten powder layer in 50 μm powder bed thickness	2012	[25]
Cu, tungsten carbide (WC) composite powder	CO_2, CW	0.6–0.9	0.3	—	FVM, Fluent 6.3.26	At 17.5 kJ m^{-1} linear energy density (LED) 96% density was achieved. The melt pool sizes are in hundreds of micrometres and increase with LED	2014	[28]
Al–1.5 Fe	Yb fibre, CW	0.6	0.3	40–80	FEM, ANSYS	The depth and width of melt pool increases with decreasing scanning speed linearly. Molten pool loses its regular form at lower than 30 mm s^{-1}	2011	[29]

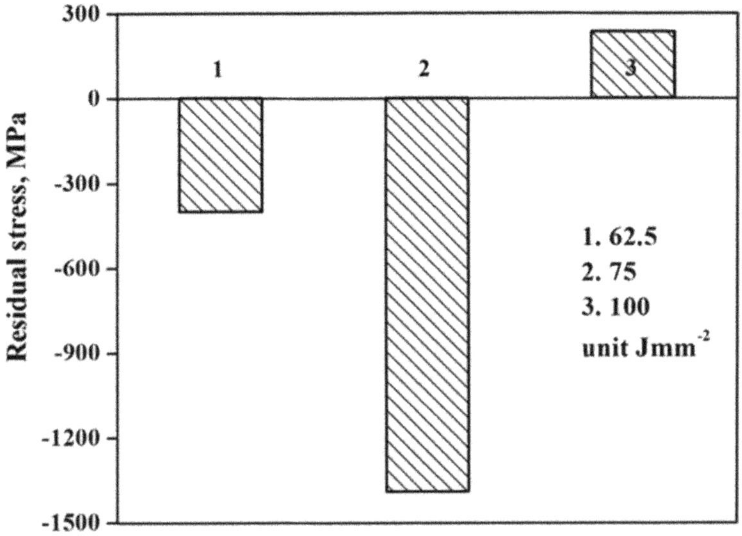

Figure 7.1. Variation in residual stresses with laser energy density; when a surface experiences melting,the compressive residual stress transforms to tensile [32].

higher in the process operated in pulse laser mode than in continuous wave mode, at equal 133 J mm^{-1} laser energy density. In addition, the increment of compressive residual stress for the pulse mode differs for pulse frequency and pulse duration [31]. On the other hand, in laser surface melting of H13 tool steel, tensile residual stress was observed and measured which can reduce the fatigue strength and toughness resulting into crack formation of the treated part at 100 J mm^{-2} energy density after melting occurred. Figure 7.1 presents the variation in type and magnitude of the average surface stress measured along the scanning direction. It is reported that at lower laser energy the surface stress was compressive due to the phase change effect. However, at higher energy the tensile stress has been developed due to the solid to liquid phase change [32]. However, there is an exception of the magnitude of stress on the surface at lower energy, because it also depends on other factors including the type of materials and nature of constraints. Through thermomechanical modelling of those processes, it is easier to predict the nature and magnitude of the residual stresses. Moreover, the mechanism of residual stress development or tendency of crack formation can also be studied through this modelling. The optimisation of process parameters in minimising induced residual stress during laser melting processes can also be carried out efficiently and effectively.

7.1.2.1 Thermomechanical modelling of laser melting processes

For the last few years, several works on developing thermomechanical models [17, 33–35] of laser melting processes were carried out to predict the residual stresses for different metals and alloys. This section will review some of the recent numerical thermomechanical models of LW, laser additive layer manufacturing (ALM), laser engineered net-shaping (LENS), and LC processes. For example, a

3D FEM thermomechanical model of hybrid laser-GTAW (gas tungsten arc welding) welding process of AISI 1018 steel plates was carried out using ANSYS. The model predicted tensile normal stress in the fusion zone and compressive in the HAZ [17]. LC of stellite (57% Cobalt 28–38% Chromium alloy) powder on a stainless steel substrate was conducted with a CW CO_2 laser. The thermomechanical modelling of this process predicted compressive residual stress in the surface of the clad and tensile stress in just below the surface spreading up to the interface [36]. Another modelling work of LC also reported a similar pattern of residual stress found in clad, interface and substrate [37]. In other models, the effects of laser parameters such as scanning speed and laser power on residual stress were also analysed. For instance, the LSM process of 42Cr4Mo steel operated by a wide band laser with rectangular heat source, and the effect of laser scanning speed on stress fields were evaluated [38]. The model predicted that at low scanning speed compressive residual stress was observed on the surface. However, at higher scanning speed the stress component on the surface turned tensile in nature. Then, laser melting in ALM of Ti6Al4V alloy powder was modelled using an ANSYS FEM solver. This model predicted compressive residual stress on the surface layer and correlated it with scanning speed. It was reported that a decrease in scanning speed increased the longitudinal stress of the Ti6Al4V surface layer [34]. The stress component along the direction of laser scanning is known as the longitudinal stress. 3D thermomechanical models of LENS processes for two different materials were also reviewed. The LENS process also lies in the additive manufacturing process family. In this process, computer-aided designed parts are built up in layer-by-layer scanning over the powder metals, similar to selective laser melting or rapid prototyping. Those models revealed compressive stress in the middle of the surface layer, while tensile stress at the free-ends [39, 40]. Important technical information including treated materials, laser parameters and the software used to develop those FEM thermomechanical models of laser melting processes have been listed in table 7.4.

All those models were developed in a general ANSYS, ABAQUS and SYSWELD commercial finite element solver using coupling techniques between thermal and structural/mechanical analyses. There are usually two kinds of coupling techniques as shown in figure 7.2, followed in FEM thermomechanical modelling of the laser processes. Where the results of thermal analysis as thermal loads due to temperature gradient are input in a structural analysis to calculate stresses induced by those thermal loads with the coefficient of thermal expansion or contraction of materials, this is known as load transfer or sequential coupling. Most of the works in table 7.5 used load transfer or a sequential coupling method to calculate residual stress. However, a direct coupling technique based thermomechanical model of laser cladding was also observed [37]. In direct coupling, both thermal and structural analyses are simultaneously conducted by applying both thermal and mechanical boundary conditions together.

In those models, the residual stress due to the temperature gradient was calculated from total strain using a classical thermo-elastic-plastic principle. The total strain, ε^{total}, consists of mainly three parts as in, $\varepsilon^{total} = \varepsilon^{th} + \varepsilon^{el} + \varepsilon^{pl}$, where ε^{th} is thermal strain, ε^{el} is elastic strain and ε^{pl} is plastic strain. The thermal strain is calculated from the temperature-dependent coefficient of thermal expansion or

Table 7.4. Materials, software and parameters used in FEM thermomechanical models of some laser melting processes.

Process name	Materials	Laser type	Laser parameters			Software	Ref.
			Power (kW)	Beam width (mm)	Scanning speed (mm s^{-1})		
LW	AISI 1018 steel	Fibre	2.5–3	0.6	25	ANSYS	[17]
ALM	Ti6Al4V powder	Nd:YAG	0.12	0.01	220	ANSYS	[34]
LENS	AISI 410 steel	Nd:YAG	$0.254–0.344_{avg.}$	0.5–0.1	2.4–8.5	SYSWELD	[39]
LENS	H13 tool steel, base-Cu/H13	Nd:YAG	—	—	2–8	SYSWELD	[40]
LC	C/S[1]-Stellite/SS[2]	CW[3] Nd:YAG	1.5	2	5	ANSYS	[36]
LC	C/S[1]-Crucible steel/H13	CW CO_2	2–3.8	3	200	ABAQUS	[37]
LSM	42Cr4Mo steel	CW CO_2	3.5	10 × 1[4]	2.4–8.5	SYSWELD	[38]
LSM	Al-10WC composite	Pulsed CO_2	1.5_{peak}	0.3	100	ABAQUS	[35]

1. C/S = clad/substrate, 2. SS = stainless steel, 3. CW = continuous wave, 4. rectangular laser beam.

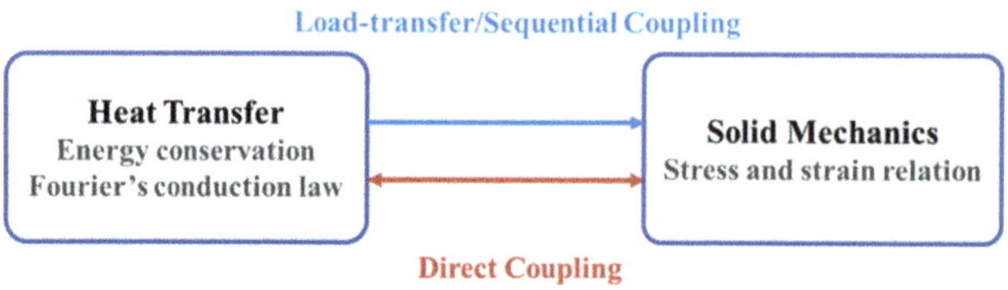

Figure 7.2. Thermomechanical coupling used in laser melting processes.

Table 7.5. The plasticity models and coupling techniques used in the thermomechanical model of laser melting processes.

Process name	Materials	Plasticity models	Coupling technique	Type of stress	Ref.
LW	AISI 1018 steel	Isotropic	Sequential	Tensile stress in the weld and compression in HAZ	[17]
ALM	Ti6Al4V powder	Isotropic	Sequential	Compressive in the surface, tensile at below the surface and at interface	[34]
LENS	AISI 410 steel	Isotropic	Sequential	Compressive in the centre of the surface and tensile at the free edges	[39]
LENS	H13 tool steel, base-Cu/H13	Isotropic	Sequential	Compressive in the centre of the surface and tensile at the free edges, stress differs with base material type	[40]
LC	C/S[1]-Stellite/SS[2]	Isotropic	Sequential	Compressive-tensile-compressive at surface-interface-substrate	[36]
LC	C/S[1]-Crucible steel/H13	Kinematic	Direct coupled	Compressive-tensile-compressive at surface-interface-substrate	[37]
LSM	42Cr4Mo steel	Kinematic	Sequential	Compressive at the surface, becomes tensile at higher scanning speed	[38]
LSM	Al-10WC composite	Isotropic	Sequential	Lower von Mises stress in the centre of the circular surface and higher near the circumference due to spiral scanning strategy which caused self-annealing at the centre	[35]

1. C/S = clad/substrate, 2. SS = stainless steel.

contraction, elastic strain follows the linear Hooke's law for metals and alloys, plastic strain is calculated from yield strength, hardening or plasticity models and flow rule of the materials. Two plasticity models, isotropic and kinematic, have been observed to calculate plastic strain based on classical plasticity theory. It is observed from table 7.5 that most laser melting processes follow the isotropic plasticity model. It is worth noting that from the literature review, no numerical thermomechanical model related to laser surface glazing of metallic materials was found.

7.1.3 Miscellaneous coupled model of laser melting processes

Some other coupled models to predict surface hardness of the laser surface melting process were also found in the literature. One method of modelling surface hardness was that of coupling the thermal model with empirical equations of phase transformation [41]. The heating and cooling curves were extracted from an FEM thermal model of the process. Then, the percentage of phase fraction was calculated using those curves utilising the time–temperature–transformation (T–T–T) diagram of the treated materials following some empirical equations of phase transformation. Figure 7.3 shows a schematic T–T–T diagram and the discretization into isothermal transformation time and temperature gradient, taken from the literature. In a T–T–T diagram, the phase transformation curve profile is drawn from the measurement of the volume fraction of phase with respect to time and temperature. In this figure, T_i is the temperature of the ith step corresponding to t_i time, ΔT is the change in temperature between the ith and subsequent i+1-th time steps and y_i is the percentage volume fraction of the y-phase at ith step.

For the diffusion-controlled transformations, for example, ferrite, pearlite, austenite transformation, the Kolmogorov–Johnson–Mehl–Avrami (KJMA) equation was used, shown in equation (7.1).

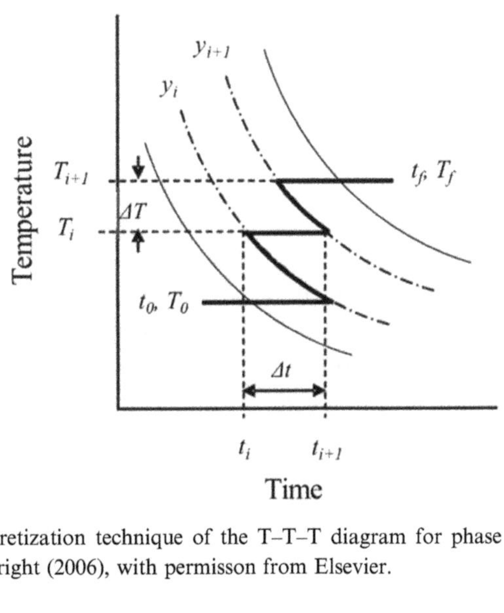

Figure 7.3. A typical discretization technique of the T–T–T diagram for phase transformation modelling. Reprinted from [41], copyright (2006), with permisson from Elsevier.

$$V_{y,\mathrm{i}} = V_y^{\max}[1 - e^{-b_k t_i^{n_k}}] \tag{7.1}$$

In equation (7.1), $V_{y,\mathrm{i}}$ is the volume fraction of the y-phase at each ith step related to isothermal transformation time t_i. $V_y^{\max}$ is the maximum volume fraction of the phases that can be formed under the considered cooling rate. This equation is used for both heating and cooling stage phase transformation. On the other hand, the Koistinen–Marburger equation was used for the displacive transformation such as martensite phase transformation [22, 42]. Equation (7.2) is used to calculate the phase fraction from the austenite to martensite phases.

$$V_{\mathrm{marten}} = V_{\mathrm{aus}}[1 - e^{-\gamma(M_s - T_{\mathrm{end}})}], \tag{7.2}$$

where V_{marten} is the volume fraction of martensite, V_{aus} the volume fraction of austenite at martensitic start transformation temperature M_s, T_{end} is the end temperature. The coefficient γ depends on the composition of steel or other alloys. Then total hardness is calculated using rule of mixture for the corresponding phase fractions present in the laser-treated zone. For calculating hardness of tempered martensite, another method called the thermo-kinetic model was followed in isothermal conditions, which is proposed by Reti *et al* [43] followed in the modelling of some other laser processes [21, 42].

7.1.4 Controlling factors in the modelling of laser melting processes

From the above literature review, it is observed that modelling of laser melting processes simplifies the physical phenomena through assumptions to find out the important factors which influence the process outcome. The most important factors which regulate the thermomechanical models of laser melting processes are materials properties, heat source distribution, laser coupling efficiency and plasticity models. To increase the modelling accuracy, it is essential to optimise those factors relative to the processes.

7.1.4.1 Material properties

The accuracy in predicted results of laser melting modelling is strongly dependent on the material properties employed in the modelling. Material properties define the response under the application of laser heat flux in the laser melting process modelling. Specific heat, thermal conductivity, thermal expansion coefficient, Young's modulus, yield strength and Poisson's ratio are the general material properties involved in a thermomechanical model of laser melting processes. As laser melting processes are heat treatment processes, therefore the dependency of the material properties on temperature is essential to account for in modelling; especially specific heat capacity, thermal expansion coefficient, and thermal conductivity that are temperature sensitive. It was reported that the numerical discrepancies of theoretical and experimental findings of the same process are due to the complex temperature-properties relation with the equivalent process condition [44]. Some numerical and modelling works of laser processes used constant material properties [7, 45, 46], which showed very distinctive prediction with the model utilising temperature-dependent material properties. It is noted that thermo-physical properties need tuning in many modelling works to conform data with the experimental works. For example, selective laser melting of Ti6Al4V powder operated with a Yb fibre laser used temperature-dependent specific heat (J kg^{-1} °C^{-1}). A Gaussian heat source was employed and 0.25 value was set for absorptivity. The depth of the melt pool was obtained 40 μm at 200 mm s^{-1} scanning speed and 110 W laser power [25]. In another Yb fibre laser-treated SLM of Ti6Al4V powder with equal scanning speed, 40 μm melt depth at 80 W laser power was achieved, presumably due to input specific enthalpy (J kg^{-1}) instead of specific heat [47]. This kind of adjustment of specific heat is also observed in laser cladding of Ti6Al4V alloy to achieve correct melt pool dimension conforming with the experimental data [20].

Therefore, employing the correct set of the temperature-dependent material properties is prerequisite to ensuring accuracy in the modelling of laser melting processes.

7.1.4.2 Heat source model

In laser melting processes, the intensity of the laser beam is transformed into thermal energy while irradiating the surface of the treated material. This thermal energy enters into the specimen as heat flux density or power density of the absorbed laser beam. The power density is defined as the amount of thermal energy flow per unit surface area at unit time. The amount of thermal energy is decided by the spatial distribution of laser intensity on the surface or beam profile. Depending on the beam optics different kinds of beam profile can be created. For example, spherical, elliptical, Gaussian and top-hat are the most common in laser melting processes [48]. Imagery software is used to determine the beam intensity profiles. In the literature, most of the modelling work of laser melting processes used Gaussian or a combination of Gaussian beam profiles such as TEM_{00} [1, 15], TEM_{01} or a combination of them [41], as shown in figure 7.4.

In the case of the Gaussian, the intensity distribution is symmetric along the beam radius and in laser melting the process symmetry is kept along the direction of the scanning laser beam. In laser welding, double ellipsoidal Gaussian distribution of

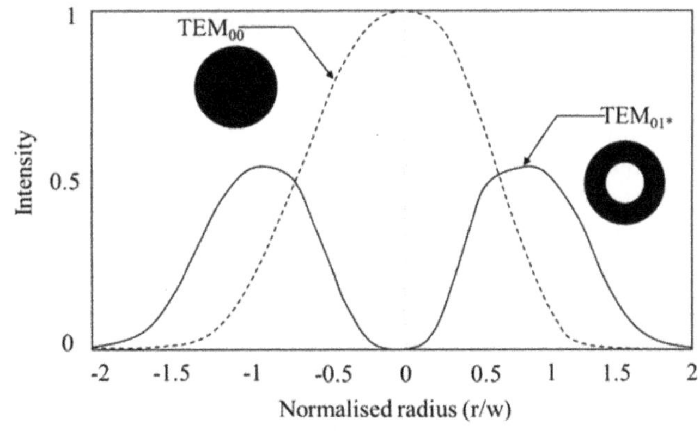

Figure 7.4. The transverse electrical mode (TEM) profile of intensity distribution of laser beam [5].

volumetric heat load is suitable [15], since Goldak introduced it for arc welding modelling [49]. A little modification in Goldak's ellipsoidal profile turned into Gaussian spherical distributed volumetric load being considered in the laser cladding process to attain the correct shape of melt pool [20]. The distribution of laser intensity is a key factor for correct prediction of the thermal profile, not only magnitude of the temperature distribution but also the correct shape of the molten and modified zone. The mathematical equations for intensity distribution of various laser heat sources have been listed in table 7.6. It is observed that most of the models utilised Gaussian distribution of laser intensity as a volumetric and surface heat source. A volumetric heat source consists of a three-dimensional distribution of laser intensity whereas a surface heat source distributes in two dimensions. However, a uniform top-hat distribution was employed in a laser melting process using a rectangular heat source [21], which was claimed to give a good metallurgical bond, low porosity and minimum dimensional distortion [3]. The Gaussian beam profile gives more penetration as the intensity concentrates more towards the centre of the beam whereas the intensity of the top-hat beam profile is more uniform throughout the beam diameter [50]. Moreover, a volumetric conical heat source with Gaussian distribution was also used in laser engineered net-shaping. The volumetric conical distribution represents maximum penetration of the laser beam where it is required [39]. The parameters related to the laser beam profile used in those equations vary depending on the laser mode such as continuous wave or pulse. Therefore, it is essential to calibrate the heat source distribution and select correctly for the process modelling to ensure accuracy of the predicted temperature, melt pool profile and residual stress distribution.

7.1.4.3 Laser coupling efficiency or absorptivity

Laser coupling efficiency or absorptivity is another crucial factor in modelling laser processes effectively. It is the ratio of absorbed laser energy by the material to the total energy provided by the laser beam, where no transmission of light through the materials is presumed [41]. In fact, the absorptivity accounts for the loss of laser intensity due to reflection or scattering of light [51] as most of the metals reflect laser light [52]. Absorptivity of laser light mainly depends on the wavelength of light, temperature, surface roughness and composition of the material [53, 54]. Absorptivity of some common metals with various wavelengths has been presented in figure 7.5.

It is noted that Fe and steel can absorb laser light no more than 10–12% for a CO_2 laser, which is in the far-infrared (10.6 μm) zone. All other metals (Al, Cu, Mo, Au, Ag) absorb very little in this wavelength. Figure 7.5 shows up to 30% absorptivity, because any as-received metals' surface hardly absorbs the laser above that range in the visible to infrared region. However, with pre-treatments such as sand blasting, chemical etching or painting of those metals' surfaces, the absorptivity can be enhanced by up to 80% [6, 56]. Different values of laser absorptivity for various metals and laser types are listed in table 7.7.

From table 7.7, it is observed that a different absorption coefficient of Ti6Al4V alloy is variably used in three distinct processes, laser assisted machining, selective laser melting and laser cladding. An experimental work calculating the absorptivity

Table 7.6. Various heat source models applied in the modelling of different laser melting processes.

Processes	Laser type	Beam width (mm)	Heat source model	Equations of heat flux density (q)	Peak temperature (°C)	Ref.
Laser engineered net-shaping	Nd:YAG	0.5–1	Gaussian volumetric conical	$\frac{2P}{\pi r_b^2}\left(1-\frac{z}{H}\right)\exp\left(1-\left(\frac{r}{r_b}\right)^2\right)$	1800–2200	[39]
Laser cladding	HPDDL	12×1[1]	Top-hat uniform	$\frac{AP}{w\times 1}$	1729	[21]
Selective laser melting	Yb fibre	0.035	Gaussian heat flux density	$\frac{2AP}{\pi r_b^2}\exp\left(\frac{-2r^2}{r_b^2}\right)$	6000	[25]
Laser cladding	Nd:YAG	1.8	Gaussian volumetric spherical	$\frac{6AP}{\pi\sqrt{\pi}r_b^3}\exp\left(\frac{-3r^2}{r_b^2}\right)$	2810	[20]
Multi-pulse laser melting	Nd:YAG	0.4	Gaussian pulse heat flux density	$A\frac{E}{\tau\pi r_b^2}\exp\left(-\left(\frac{x-(x+vt)}{r_b}\right)^2\right)$	2750	[48]

1. Rectangular beam.

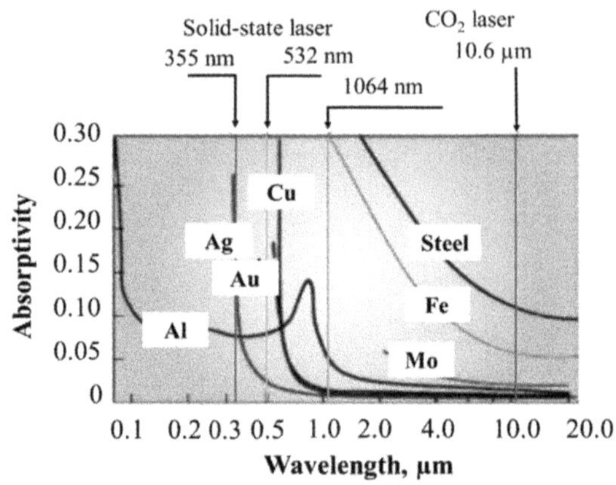

Figure 7.5. Absorptivity of some metals at different wavelengths [55]. Copyright (2013) John Wiley & Sons.

Table 7.7. Variation of absorptivity depending on laser and material types in different laser surface modification processes.

Process	Laser type	Materials	Absorptivity	References
Laser cladding	HPDDL (808 nm)	C + S[1]-H13 steel + mild steel	0.75	[21]
Laser cladding	HPDDL (975 nm)	C + S-H13 steel + mild steel	0.30	[3, 22]
Laser assisted machining	Nd:YAG (1064 nm)	Ti6Al4V plate	0.34	[57]
Selective laser melting	Yb fibre laser (1064–1100 nm)	Ti6Al4V powder	0.25	[25]
Laser cladding	CO_2 (10 600 nm)	C + S-TiC/NiCrBSiC + Ti6Al4V	0.30	[51]

1. C + S = clad + substrate.

of a Ti6Al4V plate surface revealed that the value of absorptivity varied between 0.28 and 0.4 with temperatures up to 1400 °C. Therefore, an averaged value 0.34 was employed in the subsequent modelling work of laser assisted machining of a Ti6Al4V alloy [57]. Additionally, another experimental and modelling work reported 0.45 absorptivity of a Ti6Al4V flat-surface and increased to 0.71 for powder [58]. Although, in this modelling work [25] absorptivity for the Ti6Al4V powder was adopted at 0.25 to get an accurate depth of the laser melted zone. This kind of deviation in absorptivity of the same material is observed in other works [3, 21], because of its complex nature of dependency with various factors. It is worth noting that this absorption factor also plays a key role in deciding the size and shape of the laser modified zone and enhancing the accuracy of the laser process modelling [21, 52, 59].

7.1.4.4 Plasticity models

Theoretical prediction of thermally induced residual stress through thermomechanical modelling of laser melting processes follows the classical thermo-elasto-plastic constitutive law. Here, the total strain consists of thermal, elastic and plastic strain. In metals and alloys, yield strength is evolved due to plastic strain, which eventually is added to the residual part. Similarly, in the laser surface melting process, plastic strain is caused due to the thermal load generated in the surface. The increment of this plastic strain can be calculated following three possible principles, (i) isotropic, (ii) kinematic and (iii) mixed plasticity models, based on a material's response [34]. Therefore, it is important to select an appropriate plasticity theory to correctly calculate residual stress from the thermomechanical models. In the literature, separate use of both isotropic and kinematic models has been observed. For example, Yilbas *et al* [35] modelled residual stress using the ABAQUS FEM code of laser surface melting with an aluminium composite. He considered the work-piece as an elastic body and used isotropic hardening to model rate-independent plasticity with temperature-dependent yield strength. In another work of laser cladding by Paul *et al* [60], the ABAQUS FEA software was used to model residual stress in H13 tool steel clad and substrate. A kinematic plasticity model with von Mises yield criterion was used to calculate plastic strain in this model. Although the prediction values gave rise to maximum 25% error with experimental data, the trend in stress distribution was in agreement. Li *et al* also followed the kinematic hardening rule in predicting stress distribution of laser surface melting of 42CrMo4 steel in the SYSWELD commercial FEA code [38]. It is observed that in most laser melting processes, including laser welding, laser melting, LENS and laser cladding, isotropic plasticity models are chosen for predicting thermo-elastic-plastic residual stress (see table 7.5).

7.2 Summary

A detailed literature review on recent numerical modelling works of laser surface melting processes has been conducted and presented in this chapter. A significant number of modelling works in laser welding and laser cladding has been observed followed by selective laser melting, additive manufacturing and laser surface melting. Thermal models of those processes generally predicted temperature distribution, dimension of fusion and heat-affected zones and were utilised to optimise the process parameters including laser power and scanning speed. There are also several thermomechanical models of different laser melting processes predicting residual stress in the laser modified surface. From the literature review it is evident that a single 2D numerical thermal model of LSG for metallic materials has been found, which is insufficient to grasp the process mechanism and collect the important criteria for developing an FEM model of LSG for metals and alloys. Nevertheless, the information collected from the developed FEM models of other laser melting processes can be used as a guideline for developing the LSG model. The lack of work in LSG modelling also requires the initiative of developing a reliable and simple numerical model of LSG from scratch. Therefore, in this current study, a numerical thermal model of LSG using FEM will be developed which can quantitatively predict temperature distributions, heating cooling rates and depth of modified zone for different ranges of laser parameters and metallic materials. Thermal models could be extended to the thermomechanical model to predict thermally induced residual stress in the surface and sub-surface region and crack formation tendency in the LSG process. This model can be used for optimising laser parameters such as laser power and residence time for different materials.

References

[1] Akbari M, Saedodin S, Toghraie D, Shoja-Razavi R and Kowsari F 2014 Experimental and numerical investigation of temperature distribution and melt pool geometry during pulsed laser welding of Ti6Al4V alloy *Opt. Laser Technol.* **59** 52–9

[2] Chehrghani A, Torkamany M J, Hamedi M J and Sabbaghzadeh J 2012 Numerical modeling and experimental investigation of TiC formation on titanium surface pre-coated by graphite under pulsed laser irradiation *Appl. Surf. Sci.* **258** 2068–76

[3] Farahmand P and Kovacevic R 2014 An experimental–numerical investigation of heat distribution and stress field in single- and multi-track laser cladding by a high-power direct diode laser *Opt. Laser Technol.* **63** 154–68

[4] Azizpour M, Ghoreishi M and Khorram A 2015 Numerical simulation of laser beam welding of Ti6Al4V sheet *J. Comput. Appl. Res. Mech. Eng.* **4** 145–54

[5] Mahank T A 2010 Laser glazing of metals and metallic and ceramic coatings *PhD Thesis* Pennsylvania State University, Pennsylvania

[6] Aqida S N 2011 Laser surface modification of steel *PhD Thesis* Dublin City University, Dublin, Ireland

[7] Aqida S N, Naher S and Brabazon D 2012 Thermal simulation of laser surface modification of H13 die steel *Key Eng. Mater.* **504506** 351–56

[8] Peligrad A, Zhou E, Morton D and Li L 2001 A melt depth prediction model for quality control of laser surface glazing of inhomogeneous materials *Opt. Laser Technol.* **33** 7–13

[9] Hao L and Lawrence J 2006 Melt depth prediction for laser irradiated magnesia partially stabilised zirconia *J. Mater. Process. Technol.* **180** 110–16

[10] Li J F, Li L and Stott F H 2004 Comparison of volumetric and surface heating sources in the modeling of laser melting of ceramic materials *Int. J. Heat Mass Transf.* **47** 1159–74

[11] Deng D and Kiyoshima S 2010 Numerical simulation of residual stresses induced by laser beam welding in a SUS316 stainless steel pipe with considering initial residual stress influences *Nucl. Eng. Des.* **240** 688–96

[12] Han Q, Kim D, Kim D, Lee H and Kim N 2012 Laser pulsed welding in thin sheets of Zircaloy-4 *J. Mater. Process. Technol.* **212** 1116–22

[13] Tirand G, Arvieu C, Lacoste E and Quenisset J M 2013 Control of aluminium laser welding conditions with the help of numerical modelling *J. Mater. Process. Technol.* **213** 337–48

[14] Bendaoud I, Matteï S, Cicala E, Tomashchuk I, Andrzejewski H, Sallamand P, Mathieu A and Bouchaud F 2014 The numerical simulation of heat transfer during a hybrid laser–MIG welding using equivalent heat source approach *Opt. Laser Technol.* **56** 334–42

[15] Kumar C, Das M and Biswas P 2014 A 3-D finite element analysis of transient temperature profile of laser welded Ti–6Al–4V alloy *Proc. 5th Int. and 26th All India Manufacturing Technology, Design and Research Conference (Guwahati, India)* ed U S Dixit 1–6

[16] Han Q, Kim D, Kim D, Lee H and Kim N 2012 Laser pulsed welding in thin sheets of Zircaloy-4 *J. Mater. Process. Technol.* **212** 1116–22

[17] Kong F and Kovacevic R 2010 3D finite element modeling of the thermally induced residual stress in the hybrid laser/arc welding of lap joint *J. Mater. Process. Technol.* **210** 941–50

[18] Piekarska W and Kubiak M 2013 Modeling of thermal phenomena in single laser beam and laser-arc hybrid welding processes using projection method *Appl. Math. Model.* **37** 2051–62

[19] Kumar S, Awasthi R, Viswanadham C S, Bhanumurthy K and Dey G K 2014 Thermo-metallurgical and thermo-mechanical computations for laser welded joint in 9Cr-1Mo(V, Nb) ferritic/martensitic steel *Mater. Des.* **59** 211–20

[20] Hao M and Sun Y 2013 A FEM model for simulating temperature field in coaxial laser cladding of TI6AL4V alloy using an inverse modeling approach *Int. J. Heat Mass Transf.* **64** 352–60

[21] Santhanakrishnan S, Kong F and Kovacevic R 2011 An experimentally based thermo-kinetic hardening model for high power direct diode laser cladding *J. Mater. Process. Technol.* **211** 1247–59

[22] Farahmand P, Balu P, Kong F and Kovacevic R 2015 Investigation of thermal cycle and hardness distribution in the laser cladding of AISI H13 tool steel produced by a high power direct diode laser *Proc. ASME 2013 Int. Mechanical Engineering Congress and Exposition (San Diego, California)* 1–12

[23] Hofman J T, De Lange D F, Pathiraj B and Meijer J 2011 FEM modeling and experimental verification for dilution control in laser cladding *J. Mater. Process. Technol.* **211** 187–96

[24] Louvis E, Fox P and Sutcliffe C J 2011 Selective laser melting of aluminium components *J. Mater. Process. Technol.* **211** 275–84

[25] Song B, Dong S, Liao H and Coddet C 2012 Process parameter selection for selective laser melting of Ti6Al4V based on temperature distribution simulation and experimental sintering *Int. J. Adv. Manuf. Technol.* **61** 967–74

[26] Sallica-Leva E, Jardini A L and Fogagnolo J B 2013 Microstructure and mechanical behavior of porous Ti–6Al–4V parts obtained by selective laser melting *J. Mech. Behav. Biomed. Mater.* **26** 98–108

[27] Prashanth K G, Scudino S, Klauss H J, Surreddi K B, Löber L, Wang Z, Chaubey A K, Kühn U and Eckert J 2014 Microstructure and mechanical properties of Al-12Si produced by selective laser melting: effect of heat treatment *Mater. Sci. Eng.* A **590** 153–60

[28] Dai D and Gu D 2014 Thermal behavior and densification mechanism during selective laser melting of copper matrix composites: simulation and experiments *Mater. Des.* **55** 482–91

[29] Bertelli F, Meza E S, Goulart P R, Cheung N, Riva R and Garcia A 2011 Laser remelting of Al1.5 wt%Fe alloy surfaces: numerical and experimental analyses *Opt. Lasers Eng.* **49** 490–7

[30] Kear B H, Breinan E M and Greenwald L E 1979 Laser glazing—a new process for production and control of rapidly chilled metallurgical microstructures *Met. Technol.* **6** 121–29

[31] Telasang G, Dutta Majumdar J, Padmanabham G, Tak M and Manna I 2014 Effect of laser parameters on microstructure and hardness of laser clad and tempered AISI H13 tool steel *Surf. Coatings Technol.* **258** 1108–18

[32] Telasang G, Dutta Majumdar J, Padmanabham G and Manna I 2014 Structure-property correlation in laser surface treated AISI H13 tool steel for improved mechanical properties *Mater. Sci. Eng.* A **599** 255–67

[33] Suárez A, Tobar M J, Yáñez A, Perez I, Sampedro J, Amigo V and Candel J J 2011 Modeling of phase transformations of Ti6Al4V during laser metal deposition *Phys. Procedia* **12** 666–73

[34] Roberts I A 2012 Investigation of residual stresses in the laser melting of metal powders in additive layer *PhD Thesis* University of Wolverhampton

[35] Yilbas B S, Akhtar S S, Karatas C, Ali H, Boran K, Khaled M, Al-Aqeeli N and Aleem A B J 2016 Laser treatment of aluminum composite and investigation of thermal stress field *Int. J. Adv. Manuf. Technol.* **86** 3547–61

[36] Suárez A, Amado J M, Tobar M J, Yáñez A, Fraga E and Peel M J 2010 Study of residual stresses generated inside laser cladded plates using FEM and diffraction of synchrotron radiation *Surf. Coatings Technol.* **204** 1983–88

[37] Paul S, Ashraf K and Singh R 2014 Residual stress modeling of powder injection laser surface cladding for die repair applications *ASME 2014 Int. Manufacturing Science and Engineering Conf., MSEC 2014 Collocated with the JSME 2014 Int. Conf. on Materials and Processing and the 42nd North American Manufacturing Research Conf.* vol. 2 1–8

[38] Li C, Wang Y, Zhan H, Han T, Han B and Zhao W 2010 Three-dimensional finite element analysis of temperatures and stresses in wide-band laser surface melting processing *Mater. Des.* **31** 3366–73

[39] Wang L, Felicelli S D and Pratt P 2008 Residual stresses in LENS-deposited AISI 410 stainless steel plates *Mater. Sci. Eng.* A **496** 234–41

[40] Talukdar T K, Wang L and Felicelli S D Simulation of residual stress in LENS deposited H13 tool steel on copper substrate *ASME 2011 Int. Mechanical Engineering Congress and Exposition (Denver, CO)* 1–9

[41] Tobar M J, Álvarez C, Amado J M, Ramil A, Saavedra E and Yáñez A 2006 Laser transformation hardening of a tool steel: simulation-based parameter optimization and experimental results *Surf. Coatings Technol.* **200** 6362–67

[42] Li C, Wang Y, Zhang Z, Han B and Han T 2010 Influence of overlapping ratio on hardness and residual stress distributions in multi-track laser surface melting roller steel *Opt. Lasers Eng.* **48** 1224–30

[43] Réti T, Gergely M and Tardy P 1987 Mathematical treatment of non-isothermal transformations *Mater. Sci. Technol.* **3** 365–71

[44] Yadroitsev I, Krakhmalev P and Yadroitsava I 2014 Selective laser melting of Ti6Al4V alloy for biomedical applications: temperature monitoring and microstructural evolution *J. Alloys Compd.* **583** 404–9

[45] Yilbas B S, Sami M and AbuAlHamayel H I 1998 3-Dimensional modeling of laser repetitive pulse heating: a phase change and a moving heat source considerations *Appl. Surf. Sci.* **134** 159–78

[46] El Cheikh H, Courant B, Hascoët J Y and Guillén R 2012 Prediction and analytical description of the single laser track geometry in direct laser fabrication from process parameters and energy balance reasoning *J. Mater. Process. Technol.* **212** 1832–39

[47] Verhaeghe F, Craeghs T, Heulens J and Pandelaers L 2009 A pragmatic model for selective laser melting with evaporation *Acta Mater.* **57** 6006–12

[48] Laazizi A, Courant B, Jacquemin F and Andrzejewski H 2011 Applied multi-pulsed laser in surface treatment and numerical–experimental analysis *Opt. Laser Technol.* **43** 1257–63

[49] Goldak J, Chakravarti A and Bibby M 1984 A new finite element model for welding heat sources *Metall. Trans.* B **15** 299–305

[50] Cottam R and Brandt M 2010 Development of a processing window for the transformation hardening of nickel–aluminium–bronze *Mater. Sci. Forum* **654** 1916–19 56

[51] Lei Y, Sun R, Tang Y and Niu W 2012 Numerical simulation of temperature distribution and TiC growth kinetics for high power laser clad TiC/NiCrBSiC composite coatings *Opt. Laser Technol.* **44** 1141–47

[52] Vollertsen F, Partes K and Meijer J 2005 State of the art of laser hardening and cladding *Proc. Third Int. WLT-Conference on Lasers in Manufacturing* ed E Beyer 1–25

[53] Kaplan A F H 2014 Laser absorptivity on wavy molten metal surfaces: categorization of different metals and wavelengths *J. Laser Appl.* **26** 012007

[54] Schneider M, Berthe L, Fabbro R and Muller M 2008 Measurement of laser absorptivity for operating parameters characteristic of laser drilling regime *J. Phys. D: Appl. Phys.* **41** 6

[55] Naeem M 2013 Laser processing of reflective materials *Laser Tech. J.* **10** 18–20

[56] Chikarakara E 2012 Laser surface modifications of biomedical alloys *PhD Thesis* Dublin City University, Dublin

[57] Yang J, Sun S, Brandt M and Yan W 2010 Experimental investigation and 3D finite element prediction of the heat affected zone during laser assisted machining of Ti6Al4V alloy *J. Mater. Process. Technol.* **210** 2215–22

[58] Boley C D, Mitchell S C, Rubenchik A M and Wu S S Q 2016 Metal powder absorptivity: modeling and experiment *Appl. Opt.* **55** 6496

[59] Lawrence J, Johnston E P and Li L 2000 Determination of absorption length of CO_2 and high-power diode laser radiation for ordinary Portland cement *J. Phys. D: Appl. Phys.* **33** 945–7

[60] Wang S, Singh R and Yan W 2014 Thermo-mechanical modelling of laser cladding of CPM9V on H13 tool steel *Proc. 5th Int. and 26th All India Manufacturing Technology, Design and Research Conference (Guwahati, Assam)* ed U S Dixit, R Ganesh Narayanan and M Ravi Sankar 10–6

IOP Publishing

Laser Micro- and Nano-Scale Processing
Fundamentals and applications
Ahmed Issa and Dermot Brabazon

Chapter 8

Pulsed laser ablation in liquid (PLAL) for nanoparticle generation

Brian Freeland, Eanna McCarthy, Sithara Sreenilayam, Greg Foley and Dermot Brabazon

Nanoparticles, broadly spherical pieces of material with diameters in the nanoscale range, have a number of advantageous physical, chemical, electrical, and optical properties. These unique properties make them suitable for a wide range of applications including sensing, medical therapeutics, printed electronics, and anti-fouling/anti-microbial surfaces. Pulsed laser ablation in liquid (PLAL), also known as laser ablation synthesis in solution (LASIS), is an attractive, green method for producing ligand-free nanoparticles in solution. These nanoparticles can be produced from a wide range of target materials and do not require the use of hazardous, environmentally unfriendly chemicals. In this chapter, the key applications, conventional generation methods of nanoparticles, as well as the background and cutting edge of PLAL are reviewed.

8.1 Introduction

Nanomaterials are defined as objects with one or more lengths in the nanoscale in the range of 1–100 nm [1]. Nanomaterials may have a single nanoscale dimension (thin films or sheets), two nanoscale dimensions (nanowires or nanotubes), or three nanoscale dimensions (nanoparticles). Within the dimension range of 1–100 nm, nanoparticles can range from small clusters of atoms to larger particles [1].

The unique chemical and biological functionality, high surface areas, and electrical, mechanical, and optical properties of nanomaterials make them highly suited to a number of applications such as sensing [1–3], separation [4], energy storage [5], display technology [6], printable circuits [7], and drug delivery [8]. Nanomaterials have high surface-to-volume ratios, making them well-suited to many sensing applications [1–3]. For example, in a fully one-dimensional single-walled carbon nanotube, every atom is at the surface giving it an ideal surface-to-volume ratio [2]. The relationship between

doi:10.1088/978-0-7503-1683-5ch8

the percentage of atoms at or near the surface with the size of a nanoparticle is exponentially decreasing with a particle diameter of 5 nm containing 60% volume fraction within 0.5 nm of the surface, a particle diameter of 20 nm containing 12% volume fraction within 0.5 nm of the surface and a particle diameter of 100 nm presenting just 6% volume fraction within 0.5 nm of the surface [1].

Nanomaterials also have attractive electronic properties. When dimensions are greater than the mean free path of a carrier within the material, a nanoscale object will have the same electron transport properties as the bulk material. However, when one or more of the dimensions are smaller than this, the electronic properties become dependent on the dimension(s) [9]. This allows electronic properties, such as voltage outputs, turn on/off currents, and characteristic sensor responses, to be tuned by control of the nanoscale dimensions [10–12]. The optical properties may also be varied with control of the shape and dimensions. Dreaden *et al* presented different photon absorption behaviour (due to collective conduction electron excitation or plasmon resonance) with different sizes and shapes of nanomaterials, identifiable by the differing visual colours [13].

Pulsed laser ablation in liquid (PLAL), also known as laser ablation synthesis in solution (LASIS), is a method of nanomaterial fabrication where a pulsed laser is focused on a solid target in a liquid medium, ablating the material to form nanomaterials in solution [14]. This method allows for production of nanoparticles using various target materials, without the need for environmentally hazardous solvents. However, commercially viable mass production with this method has been limited by batch-based production leading to low outputs.

In this chapter, the key applications of nanoparticles will be reviewed, a general overview of nanoparticle generation and a more detailed review of PLAL nanoparticle generation will be presented, including the current state-of-art for PLAL and issues facing commercialisation of the process.

8.2 Nanoparticle applications

8.2.1 Sensing

The high surface areas and optical and electrical properties of nanoparticles make them ideal for sensing applications. Most nanoparticle-based sensors can be broken down into colorimetric, fluorescence, or electrochemical sensors [15]. The photon absorption behaviour, and thus the visual colour, of nanoparticles is strongly dependent on the size and shape of the particles [13]. These colourful light-scattering properties allow nanoparticles to be used similarly to fluorescent dyes. The light-scattering power of a single nanoparticle label is stronger than a single fluorescent label, and the light signals are not subject to photobleaching and require less complex instrumentation [16].

Typically, colorimetric sensing uses larger nanoparticles (>30 nm), which exhibit strong visible light scattering. Gold nanoparticles are particularly popular, as they have a surface plasmon resonance, where conduction electrons near the surface of the metal are stimulated into oscillation by an external electromagnetic wave, with a maximum absorption at a resonance frequency. For nanoparticles, the wavelength of this resonance is strongly dependent on the size, shape, aggregation, and medium refractive index [17]. Changing these parameters will change the visual colour of the nanoparticles. For example, well-dispersed gold nanoparticles with sizes of 10–50 nm will appear red, while aggregation of the particles will change the colour to blue [18–20]. This provides a mechanism for colorimetric sensing.

A typical aggregation-based sensing process was detailed in a work by Chen *et al*, describing colorimetric detection of melamine in milk using gold nanoparticles [21]. Melamine is an industrial compound used in manufacturing that can be used to adulterate milk to create false high protein values. Excessive melamine intake is harmful, especially to infants and adolescents, driving the need for reliable methods of detecting melamine in milk [22]. The gold nanoparticles were chemically synthesised by a trisodium citrate reduction, then chemically modified to asymmetrically graft polyethylene glycol (PEG) as a stabiliser. The presence of melamine in the milk disturbs the electrostatic balance on the surface of the nanoparticles, promoting the formation of aggregates, shown in figure 8.1. This results in a colour shift, as described above, which can be detected by eye or with spectrophotometry. The authors found this method to be sensitive and reliable. Methods may also be based on 'anti-aggregation', where the target prevents or impedes aggregation of the nanoparticles [15]. An example of an anti-aggregation method is presented by Ramezani *et al*, in a work where gold nanoparticles were used in colorimetric sensing of tetracycline [23]. Tetracyclines are broad-spectrum antibiotics used in veterinary medicine. Residues of tetracyclines in food products could have undesirable side effects on human consumers, creating a need for tetracycline sensing. Gold nanoparticles were chemically synthesised by a citrate reduction and combined with a triple helix molecular switch (THMS) made up of a purchased aptamer and signal transduction probe (STP). THMS is stable in the absence of tetracycline allowing ambient salt (NaCl) in the medium (serum or milk in this study) to trigger aggregation of the nanoparticles, leading to a colour change to blue [23]. When the targeted tetracycline is present, the aptamer in the THMS

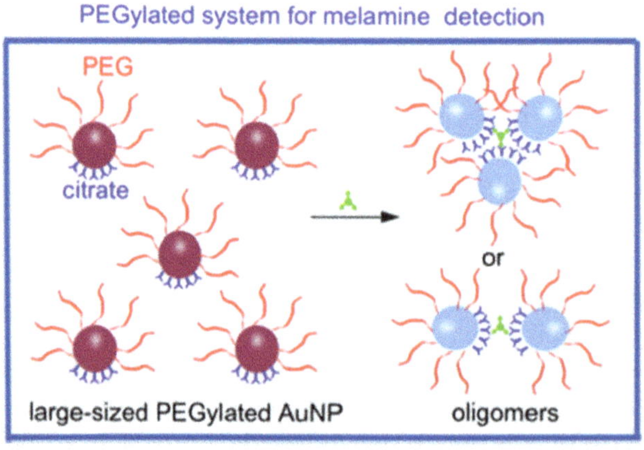

Figure 8.1. Principle of asymmetrically PEG-modified gold nanoparticle aggregation-based colorimetric sensor for detection of melamine. Reprinted from [21], copyright (2019), with permisson from Elsevier.

binds to the target, and the released STP binds to the nanoparticles inhibiting aggregation leaving the red colour. As with aggregation-based methods, the colour can be identified visually or with spectrophotometry.

Fluorescence sensing is based on the use of organic or inorganic fluorophore dyes, which suffer from some limitations. The dyes can have low absorption coefficients and weak signals, limiting sensitivity and response, are prone to photobleaching, leading to short lifetimes, and are potentially toxic [16, 24]. Fluorescent nanoparticles, made up of organic fluorophores encapsulated in a particle matrix, are much brighter than single dye molecules, and are more stable and biocompatible than un-encapsulated fluorophores [24]. Nanoparticles can also be used as quenchers in fluorescence resonance energy transfer (FRET) based sensing [25]. FRET is a highly distant dependent, non-radiative process where an excited donor fluorophore transfers energy to an acceptor [15, 25]. The initiation of FRET can act to quench or turn off fluorescence, as energy that would be otherwise radiatively released through fluorescence is transferred into the acceptor, which can act as a sensing method [15]. For example, Xu *et al* describe the use of gold nanoparticles in a FRET based sensing strategy for sensing the neurotransmitter dopamine [25]. In this work, a dopamine binding aptamer and a fluorescent (rhodamine B) were used with citrate-synthesised gold nanoparticles. When dopamine is not present, the aptamer attaches to the gold nanoparticles, preventing NaCl induced aggregation, and FRET quenching can take place between the nanoparticles and the fluorophores. When dopamine is present, the aptamer will preferentially attach to the target dopamine, allowing the nanoparticles to cluster, preventing FRET quenching. A schematic of this can be seen in figure 8.2 [25]. The authors found the method to be rapid, simple, selective, and sensitive.

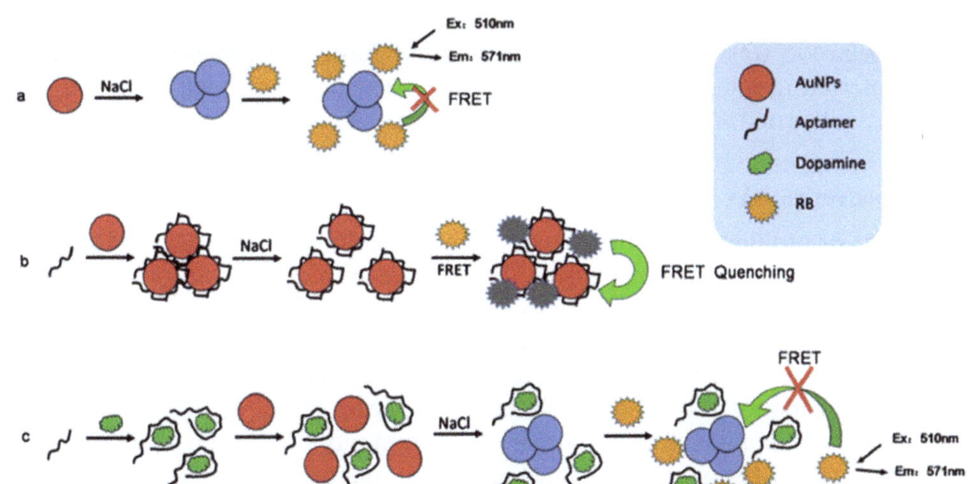

Figure 8.2. Schematic illustrating FRET quenching based dopamine sensing using gold nanoparticles (AuNPs) and rhodamine B. Reprinted from [25], copyright (2015), with permisson from Elsevier.

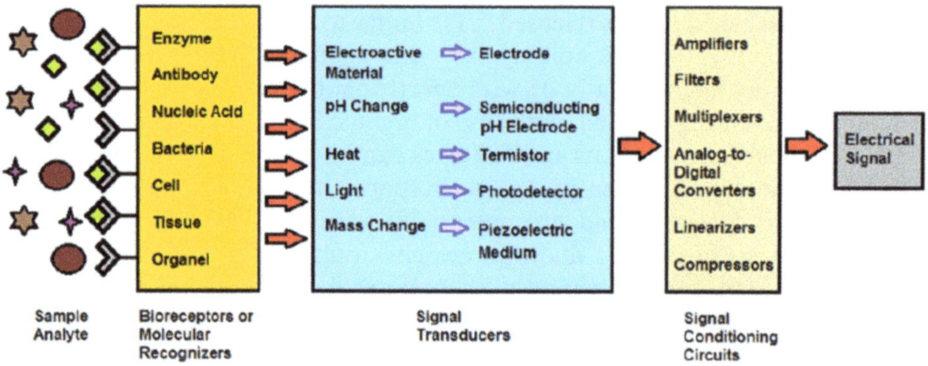

Figure 8.3. Schematic of the main subsystems of an operating biosensor. (Reproduced from [28].)

In electrochemical sensing, nanoparticles are often used to modify sensing electrodes to achieve the desired sensitivity and selectivity [15]. Electrochemical sensors work by the variation of the electric response of a device due to chemical interactions between target analytes and the surface of the electrode [26, 27]. A general schematic for an operating electrochemical biosensor is presented in figure 8.3 [28]. The high surface-to-volume ratio of nanoparticles offers a large sensing surface, and they also offer excellent electrical conductivity and biocompatibility [29]. An example of nanoparticle use in electrochemical sensing can be seen in a work by Raj *et al*, where immobilised gold nanoparticle arrays were used for voltammetric sensing of dopamine [30]. The authors note that detection of dopamine is inhibited by the presence of interfering compounds in biological samples, and as such ensuring good selectivity in sensing electrodes is a major goal. Treatment or coating of electrodes is one way to achieve this. The authors applied an amine terminated self-assembled monolayer (cystamine) to a gold electrode, then immobilised chemically synthesised gold nanoparticles on the surface. The nanoparticle coated electrode was shown to be able to distinguish between dopamine and a typical interfering compound (ascorbate), which was not possible with the bare gold electrode, with good sensitivity, selectivity, and anti-fouling.

8.2.2 Conductive inks

Deposition or printing of conductive inks is an efficient method for the production of conductive coatings and circuits [31]. Printed electronics can be low cost, light weight, optically transparent, and used in flexible electronics [32]. Conductive patterns can be printed to flexible substrates such as paper [33–35] or polymer [36, 37]. Flexible electronics are of interest for radio frequency identification (RFID) tags [33, 35], robotics [38, 39], and biomedical applications [40, 41]. Conductive inks of conductive nanomaterials in solution are commonly used, such as silver nanowires [42], silver nanoparticles [36, 43], copper nanoparticles [44], and carbon nanotubes [45]. Silver nanoparticle inks are one of the most commonly used, due to their high conductivity, and low oxidisation [46, 47]. Silver nanoparticle inks are typically in aqueous or solvent media, stabilised with surfactants or polymers, and

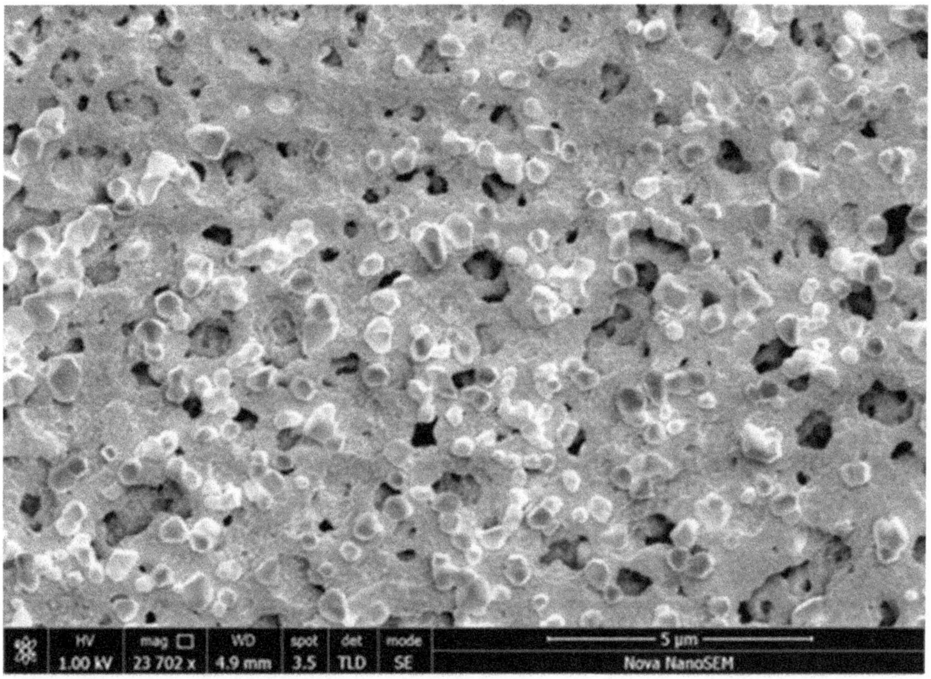

Figure 8.4. SEM image of a printed silver nanoparticle circuit on a PET substrate. (Reproduced from [36]. Copyright IOP Publishing, reproduced with permission, all rights reserved.)

then dried and/or sintered after printing [46]. An example of silver nanoparticles being used in a conductive ink can be seen in a work by Matyas *et al* [36]. In this work, silver nanoparticles were produced by a solvothermal precipitation synthesis method using silver nitrate salt, an organic precipitation agent, and a polymer stabilising agent. The ink was prepared by dispersing the nanoparticles in deionised water with ultrasonication, with 0.1 ml of dispersion stabiliser added per 10 ml of dispersion fluid. The final viscosity of the ink ranged from 8 to 12 cPs. The ink was then ink-jet printed at 35 °C onto a PET-based PCB substrate at 45 °C. After printing, the substrate and printed circuit were dried in a vacuum oven at 120 °C for 20 min to sinter the nanoparticles. An SEM image of the printed silver nanoparticle layer can be seen in figure 8.4 [36]. With this method, a working flexible antenna was successfully created.

8.2.3 Anti-fouling

Fouling is the deposition or accumulation of unwanted material on the surface of a solid. This gathering of material can inhibit the mechanical, chemical, or electrical function of devices, lead to contamination, or, in the case of medical tools or devices, lead to infection. Fouling is a significant problem for applications like sensing electrodes [48] and osmosis membranes [49]. Anti-fouling surfaces, with properties such as superhydrophobicity, self-cleaning, and drag reduction, which resist fouling are therefore highly desirable [50].

Silver and its compounds are well known for their anti-microbial, biocidal effects [51–53]. Silver nanoparticles are particularly popular for use in anti-microbial coatings due to the high surface area, allowing for more reactive surface-oriented groups compared with the bulk material [52]. Silver nanoparticles have a number of mechanisms against micro-organisms: particles attaching to the cell surfaces, disrupting the function of the cell wall [54, 55]; permeating the cell and damaging the DNA, proteins, or other cell constituents [55, 56]; and releasing reactive silver ions [55, 57]. Titanium dioxide nanoparticles also have good anti-fouling properties and wettability [58]. A work by Nguyen *et al* describes the use of silver nanoparticles and titanium dioxide nanoparticles for anti-fouling effects on a forward osmosis membrane [49]. The build-up of deposits of organic matter on osmosis membranes can reduce their permeability and shorten their lifetime. The authors note that the antibacterial effect of silver nanoparticles can be lost if the particles become covered by a biofilm. In this work silver nanoparticles were combined with titanium dioxide, which is known for its propensity to decompose organic matter, to create an improved anti-fouling surface. Silver and titanium oxide nanoparticles were deposited in separate wet chemical processes onto the porous surface of commercial forward osmosis membranes, which were tested for their anti-fouling behaviour. They found the modified membrane had excellent antibacterial effect, almost 11 times less bacterial growth compared to the as-received membrane, and that the flux reduction due to fouling was lessened for the modified membrane due to the delayed onset of fouling.

8.2.4 Therapeutics

Nanoparticles have been used in therapeutics for illnesses such as Alzheimer's [59–61], cancer [62–64], and atherosclerosis [65]. Nanoparticles may be used in imaging-based therapeutic techniques [63], drug delivery [59], or gene delivery [66]. Chemical synthesis is a common method of producing nanoparticles, however the chemicals involved in their production may be hazardous to human health, and as such alternative production methods may be more attractive for therapeutic applications [63].

The unique optical properties and the surface plasmon resonance of gold nanoparticles make them of interest for imaging-based therapeutic techniques or photo-thermal therapy [63]. Photo-thermal therapy is a cancer therapy where photon energy is converted into heat energy to destroy cancer cells [67]. Synthetic molecules which absorb the applied photons more strongly than the surrounding tissue can be applied to a tumour to localise the heating. In plasmonic photo-thermal therapy, gold nanoparticles are introduced to the cancerous cells by intravenous or intratumoural injection, and then exposed to light [67, 68]. The light, typically near infra-red, causes oscilliation of the free electrons in the gold nanoparticles due to surface plasmon resonance, and this oscillation energy can be emitted as heat through non-radiative decay. This strong localised heating then destroys the cancer cells [67].

Nanoparticles can also be a drug delivery method [59–61, 64]. As carriers or vessels for drugs, nanoparticles have the potential advantages of penetration of biological barriers like cell membranes or the blood–brain barrier, protection of the drug from premature chemical or physical breakdown, and surface modification to help with solubility and delivery or limit toxicity [59, 64, 69, 70]. Improving the specificity/targeting of drugs is also possible [59, 63].

Chemotherapy drugs, for treatment of cancer, are traditionally delivered orally or intravenously, causing them to affect the entire body [63]. The resulting negative side effects drive an interest in targeted or localised drug delivery methods. Nanoparticles are one method of achieving this. Chemotherapy drugs can be loaded or attached to nanoparticles. Tumour tissue tends to have leaky vasculature, which will cause an accumulation of intravenously applied nanoparticles from the blood stream as a form of 'passive' targeting [63]. However the strength of this targeting varies with the type of tumour due to variation in the vasculature [71]. In 'active' targeting, ligands of tumour specific biomarkers are conjugated to the nanoparticle carrier. The ligand's interaction with the tumour causes the nanoparticles to be internalised by the tumour cells by endocytosis [71]. Careful selection of the ligand is necessary, but greater targeting can be achieved for a greater range of tumours using active compared to passive targeting. Gold nanoparticles are well-suited to this application due to their biocompatibility, non-toxic nature, and tuneable sizes, geometries, and properties [72, 73]. For example, Chen *et al* describe the conjugation of a chemotherapy agent methotrexate to gold nanoparticles to treat tumours [74]. The gold nanoparticles were chemically synthesised with a citrate reduction. A colloidal solution of the nanoparticles was then mixed with methotrexate and a sodium phosphate buffer. The mixture was centrifuged and rinsed, and then redispersed in the buffer. A diagram of the conjugation can be seen in figure 8.5 [74]. A coordinate-covalent bond attaches the methotrexate molecule to the gold nanoparticle. The authors observed that the drug accumulates at a faster rate, to a higher amount, when conjugated to the gold nanoparticles compared to the drug alone being administered.

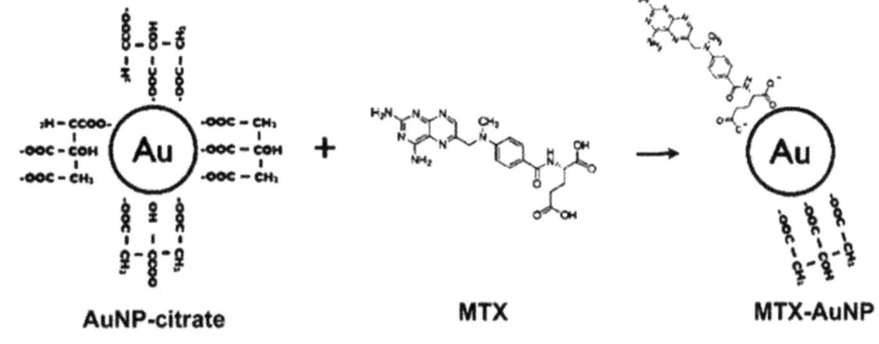

Figure 8.5. Schematic of the conjugation of methotrexate (MTX) to the citrate-reduced gold nanoparticles (AuNP). Reprinted with permission from [74]. Copyright (2007) American Chemical Society.

Diseases like Alzheimer's disease (AD), which affect the brain and central nervous system, can be difficult to treat, due to the blood–brain barrier [59]. Drugs used in AD therapy, typically orally administered, tend to have side effects due to a lack of selectivity for the therapeutic targets [60, 61]. This issue can be mitigated with drug carriers capable of crossing the blood–brain barrier, allowing sufficient amounts of the drug to be delivered to the brain and central nervous system at lower overall doses, minimising off-target effects [59]. Nanoparticle drug carriers can successfully deliver drugs through the blood–brain barrier [75]. The nanoparticles used are typically dendrimers (dendritically branched molecules), polymer nanoparticles, and lipid nanoparticles [59, 75].

8.3 Non-laser based nanoparticle generation

8.3.1 Chemical generation

Chemical synthesis is one of the most common methods of producing metal nanoparticles [76]. Reduction based methods in solution are typical for metallic and alloy nanoparticles [77]. Silver nanoparticles are most commonly produced by chemical reduction with organic or inorganic reducing agents, such as sodium citrate, ascorbate, sodium borohydride, elemental hydrogen, polyol process, Tollens reagent, *N*-dimethylformamide, and poly (ethylene glycol)-block copolymers [78]. The reducing agents reduce Ag^+ leading to the formation of metallic silver, which agglomerates into nanoparticles. Stabilising agents can be used to prevent further agglomeration of the nanoparticles into larger clusters [79]. The particles may also be functionalised. Similar chemical reduction methods are used for gold nanoparticles, using reducing agents such as borohydrides, aminoboranes, hydrazine, formaldehyde, hydroxylamine, saturated and unsaturated alcohols, citric and oxalic acids, polyols, sugars, hydrogen peroxide, sulphites, carbon monoxide, hydrogen, acetylene, and monoelectronic reducing agents [80]. As with silver, stabilising agents are added to maintain dispersion of the particles. Citrate-based reduction methods are most common for gold. Citrate can act as both a reducing and a stabilising agent for gold nanoparticles. This method is known as the Turkevich method, developed by Turkevich *et al* in 1951 [81], and then improved upon by Frens in 1973 [82]. In this method chloroauric acid ($HAuCl_4$) is boiled and trisodium citrate is then added under stirring [80]. Citrate can also be used solely as a stabilising agent, with a separate reducing agent [83].

A disadvantage of some chemical synthesis methods are their use of chemicals and solvents (such as hydrazine and sodium borohydride), which may be hazardous to humans or the environment [63, 84]. An interest in 'green'—ecologically safe—methods of generating nanoparticles has driven some movement away from chemical synthesis to other methods such as PLAL, but also in developing green synthesis methods using biological agents in the reducing process [84]. These methods are generally based on reducing metal salt solutions with a biological agent, and have been used with silver, gold, platinum, mercury, selenium, palladium, and others, including alloys and metal oxides [84–86]. A general diagram for the process, for silver nanoparticles, can be seen in figure 8.6 [84]. Biological agents may

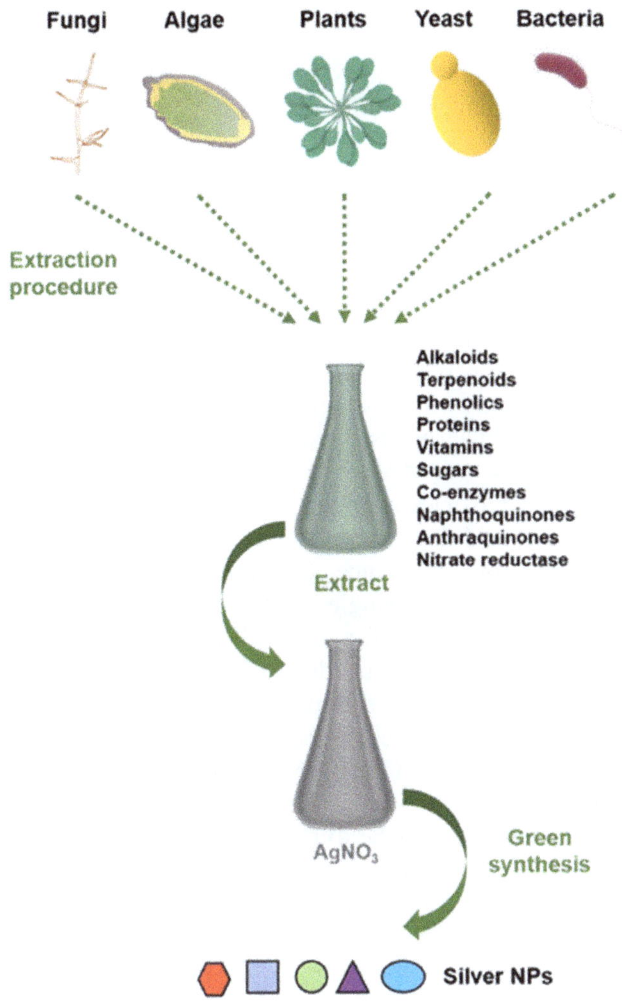

Figure 8.6. Schematic for green synthesis of silver nanoparticles using various biological agents. (Reproduced from [84] with permission of The Royal Society of Chemisty.)

be taken from plants, microbial, algal, and cyanobacterial sources, and typically the agents chosen act as both reducing and stabilising agents [84]. For example, Dwivedi and Gopal report on the production of silver and gold nanoparticles using an extract from Chenopodium album leaves [87]. Chenopodium album is a weed that is found in Asia, North America, and Europe. The authors prepared a leaf extract by boiling fresh, washed leaves in distilled water. The extract was then used as a reducing agent to produce silver and gold nanoparticles from silver nitrate and auric acid, respectively, in aqueous solution. The reaction was carried out at room temperature for 15 min. This simple, convenient, green method successfully produced 10–30 nm nanoparticles for both silver and gold.

8.3.2 Physical generation

Methods of nanoparticle generation based on physical processes can avoid the use of potentially hazardous or contaminating chemicals and solvents, and can achieve more uniform distribution of nanoparticles compared to chemical processes, however they may require higher capital costs for equipment, and higher energy consumption, for thermal or laser based processes [78]. Evaporation–condensation methods, where a source material is evaporated by heating, for example in a tube furnace, and condensed into nanoparticles are one physical generation method. Magnusson *et al* report on the production of gold nanoparticles by evaporation with a tube furnace, aerosolisation, size selection, and thermal reshaping [88]. In this method, the source material is placed in a ceramic boat in a tube furnace, heated to create a high pressure vapour, which is then carried by a nitrogen carrier gas flowed through the furnace, creating aerosolised particles. The particles are size selected in a differential mobility analyser (DMA), which relies on the particles being charged beforehand. The selected particles are then reheated in a second furnace to allow reshaping of the particles, and then size selected again by a second DMA. With this method, the authors were able to produce gold nanoparticles in the 20 nm size range, with near spherical shape and near single crystallinity.

Arc discharge has also been used for physical nanoparticle generation. Lo *et al* describe the use of arc discharge to produce silver nanoparticle suspensions [89]. A diagram of the experimental set-up is presented in figure 8.7 [89]. A silver bar submerged in a dielectric liquid is used as the electrode. An electric arc is generated, heating the metal to 6000–12 000 °C. The evaporated material quickly cools in the surrounding low temperature dielectric liquid and forms silver nanoparticles dispersed in this cooling medium. The authors report the novel process was, with the correct processing parameters, successful in producing well-dispersed nanoparticles in deionised water with average sizes ranging 6–25 nm, and avoiding

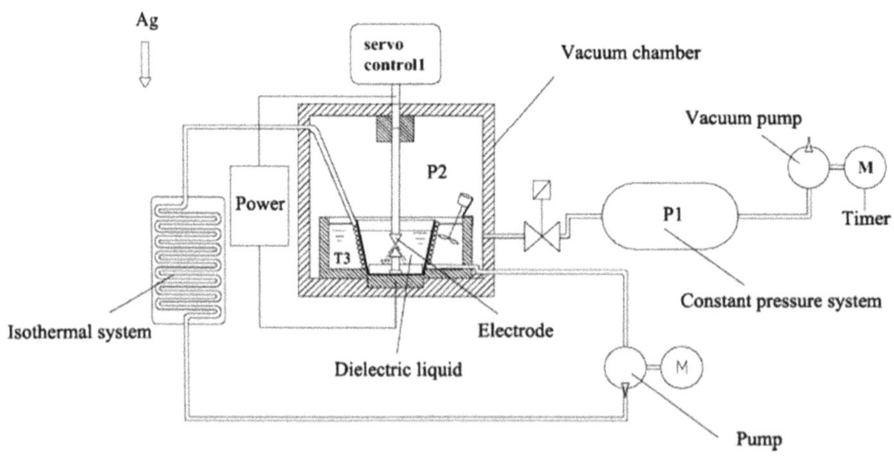

Figure 8.7. Schematic diagram of the arc discharge process for silver nanoparticle generation. Reprinted from [89], vopyright (2007), with permission from Elsevier.

aggregation. Lung *et al* describe the use of arc discharge to produce gold nanoparticles [90]. Gold wires submerged in deionised water were used as both the positive and negative electrodes. Arc discharge between the wires, as with the process described by Lo *et al* [89], evaporates the metal, which cools and condenses in the surrounding deionised water. The authors report that the method is cheap, rapid, environmentally friendly, and successfully produces gold nanoparticles, without the need for any surfactants or stabilisers.

Nanoparticles can also be produced with physical grinding or milling. Grinding methods can be divided into dry or wet grinding [91–93]. Production of nanoparticles by dry ball milling presents a simple, low cost, environmentally friendly, high yield method [91, 93]. Arbain *et al* report on the production of iron oxide nanoparticles by dry milling [93]. In this work a planetary ball mill was used under atmospheric conditions with steel grinding media in a 3:1 ratio with the raw material, which was high purity haematite (Fe_2O_3) powder. Nanoparticles were successfully produced with a minimum size of 76.6 nm, without the need for any chemical additives, however there was some aggregation of the particles.

Wet milling allows the production of suspensions, and can produce more stable nanoparticles with the use of stabilising agents, however this reduces the simplicity of the method, and may add cost and reduce the environmentally friendliness. Knieke *et al* report the production of tin oxide and aluminium oxide nanoparticles by wet grinding [94]. Aluminium oxide powder of average particle size 2 μm or tin oxide powder of average particle size 1.6 μm were used, in deionised water and denatured ethanol, with a zirconia grinding medium. The powders were premixed with the wet chemicals in a stirred vessel, pumped through a heat exchanger into the mill, and returned to the stirred vessel [94].

Gold and silver nanoparticles have also been produced by wet grinding. Pimpang *et al* report on manual grinding of silver and gold nanoparticles [92]. Silver powder and gold foil were used as raw materials, with deionised water and ethylene glycol and 5 wt% polyvinyl alcohol as medium and stabilisers. The nanoparticles were produced by hand grinding in an agate mortar. The authors successfully produced gold and silver nanoparticles with optical properties indicating particle sizes of 140–150 nm.

Grinding/milling may be used as part of a chemical production process to shape or stabilise the nanoparticles. Balaz *et al* report on a combined chemical–physical production method for silver nanoparticles, by wet milling using a plant-based reducing agent [95]. Silver nitrate and polyvinylpyrrolidone were used with an extract from *Origanum vulagare* L. plants. The leaves, flowers, and stems of *O. vulgare* L., a herb in the mint family, were dried and powdered to fine particles, which were then added by 10 g to 100 ml of distilled water. The mixture was then heated to 60 °C for 10 min, cooled down, and filtered to produce the reducing agent. The nanoparticles were produced in solution by chemical reduction with the reducing agent, then processed in a stirring media mill with a zirconium dioxide milling medium. The stabilising agent, polyvinylpyrrolidone, was added in this milling process. This method successfully produced stable silver nanoparticle suspensions, with two size groups of average size 7 and 38 nm, respectively.

8.4 Laser based nanoparticle generation

8.4.1 PLAL generation

PLAL is a physical method of nanoparticle generation where a pulsed laser is focused on a solid target in a liquid medium, to ablate material from the target, creating nanoparticles in solution [14]. The laser light irradiating the solid target is partially absorbed by the target (with the remaining light being scattered or transmitted). The energy is absorbed by the excitation of bound electrons in the target. If the absorbed energy is high enough, the bound electrons will be released, ionising the material. These free electrons can continue to absorb the laser light by inverse Bremsstrahlung, leading to further heating and ionisation [96]. The result of this is the formation of high pressure plasma, which expands out from the target in a plume, initially with high speeds, which generate a shockwave through the liquid. As the plume expands it forms a gas cavitation bubble in the liquid, which collapses under the surrounding pressure, producing another shockwave that expels the material into the liquid, cooling and condensing it into solid form, producing nanomaterials such as nanoparticle colloids. A diagram for this process is shown in figure 8.8 [97]. The collapse of the bubble may produce a rebound which then collapses, potentially producing another rebound, and so on, giving behaviour similar to a damped oscillator [98–100].

The morphology of the nanomaterials produced can be controlled by altering the process parameters such as laser fluence, wavelength, and ablation medium. The PLAL process is highly adaptable, capable of producing a variety of nanomaterials (nanoparticles, nanocubes, nanorods, etc) with a variety of compositions (metals, alloys, oxides, carbides, hydroxides, etc) [101]. The method generally doesn't require

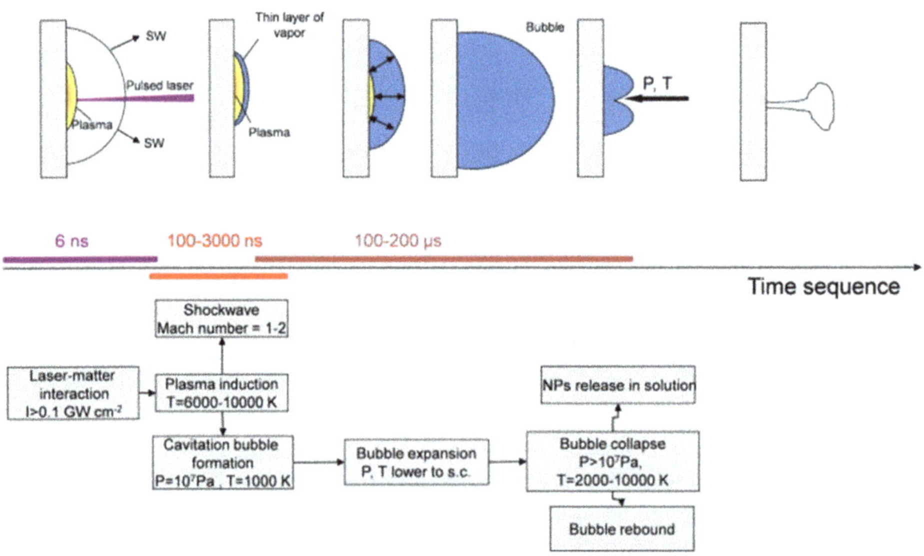

Figure 8.8. The time sequence of the processes involved in pulsed laser ablation in liquid. Reprinted from [97], Copyright (2015), with permission from Elsevier.

additional chemical agents, beyond the liquid solvent medium, making the method eco-friendly and producing surfactant and ligand-free nanoparticles. Some applications, like catalysis and biomedicine, require ligand-free nanoparticles, that would otherwise need to be obtained through cleaning of the nanoparticles [102].

There has been some investigation into the cavitation bubble and nanomaterial formation process in PLAL. Ibrahimkutty *et al* monitored the formation of nanoparticles in the cavitation bubble using high time resolution, small angle X-ray scattering [103]. Nanoparticles were produced by PLAL of a gold ribbon target in water by an Nd:YAG laser (1064 nm, 6 ns, 10 mJ, 200 Hz) focused to a 100 μm diameter spot size. The gold ribbon was continuously in motion perpendicular to the laser, to present a fresh ablation site for each pulse, with a flow of deionised water in the same direction to transport the nanoparticles away from the beam focus. The cSAXS 13.6 keV beamline at the Swiss light source was used as the X-ray source. The X-rays are transmitted through the cavitation bubble, where some of the photons will be scattered by the forming nanoparticles, to a detector [103].

The authors report detection of two species within the ablation region: small particles of 8–10 nm, and larger, agglomerated, particles of ~45 nm diameter. Such bimodal size distributions are commonly reported for PLAL nanoparticles [104–107]. The authors report that the presence of primary particles decays towards the top of the bubble, with some primary particles found outside of the bubble perimeter prior to its collapse [103]. They note that initial plasma velocities for laser plasma plumes created in vacuum have been found to be 10^8 mm s^{-1}, which is sufficient to allow injection of atomic species from the cavitation bubble into the surrounding liquid [108, 109]. Larger particles have already formed, possibly due to collisions between the primary particles in the plume/bubble. These particles are not detected close to the target, which the authors attribute to the probability of agglomeration over time and height, and the confinement and cooling of particles nearer the vapour–liquid interface.

Shih *et al* investigated the bimodal distribution which commonly occurs in PLAL by experiment and modelling [110]. Atomistic modelling of picosecond laser ablation of silver in water was carried out, and imaging of cavitation bubbles in PLAL experiments, to examine the mechanics behind the formation of the bimodal size distribution. Their model suggests the formation of a transient hot molten metal layer at the interface with the surrounding liquid, which plays a critical role in the nanoparticle formation process. The liquid in contact with the hot molten layer becomes supercritical creating a metal–water mixing region that is a precursor to the cavitation bubble. The conditions in this relatively low-density mixing region are conducive to the formation of small nanoparticles from the evaporating metal (<10 nm). The authors note a secondary formation process, where the molten layer itself may break up under the pressure exerted by the surrounding supercritical liquid, forming larger nanoparticles (10–20 nm) without evaporation. This suggests a bimodal distribution, even in the absence of agglomeration. The model indicates ejection of this material from the mixing region into the surrounding cooler liquid, which was confirmed by the imaging experiments. The authors note good agreement between the model, the imaging results, and the resulting nanoparticles (analysed by centrifuge and transverse electron microscopy), supporting the conclusion that these two formation mechanisms lead to the bimodal particle size distributions commonly observed for PLAL.

De Giacomo *et al* investigated the dynamics of PLAL cavitation bubbles during nanoparticle production using three complementary optical techniques: optical emission spectroscopy (OES) to characterise the plasma, fast shadowgraphs for the plasma and cavitation dynamics, and laser scattering for the produced materials in the liquid [96]. A diagram of the set-up can be seen in figure 8.9 [96]. An Nd:YAG laser (532 nm, 8 ns, 10 Hz) was used for ablation of a solid metal target, in the form of a wire or a solid bulk, in water. By measuring the plasma temperature using OES and the initial volume from the shadowgraph images, the authors could calculate an initial pressure of the cavitation bubble, estimating values of 10^8 Pa for metal bulk targets ablated with laser fluences of 6 J cm^{-2}. When the bubble reaches its maximum expansion, the temperature of the bubble will be approximately equal to that of the surrounding liquid, and the pressure will be equal to the saturation pressure at that temperature. This pressure will be orders of magnitude lower than the surrounding liquid pressure, causing the bubble to contract. The authors note that this process is important, as produced nanomaterials may be deposited onto the surface of the target instead of being driven out into the solution. The shadowgraph,

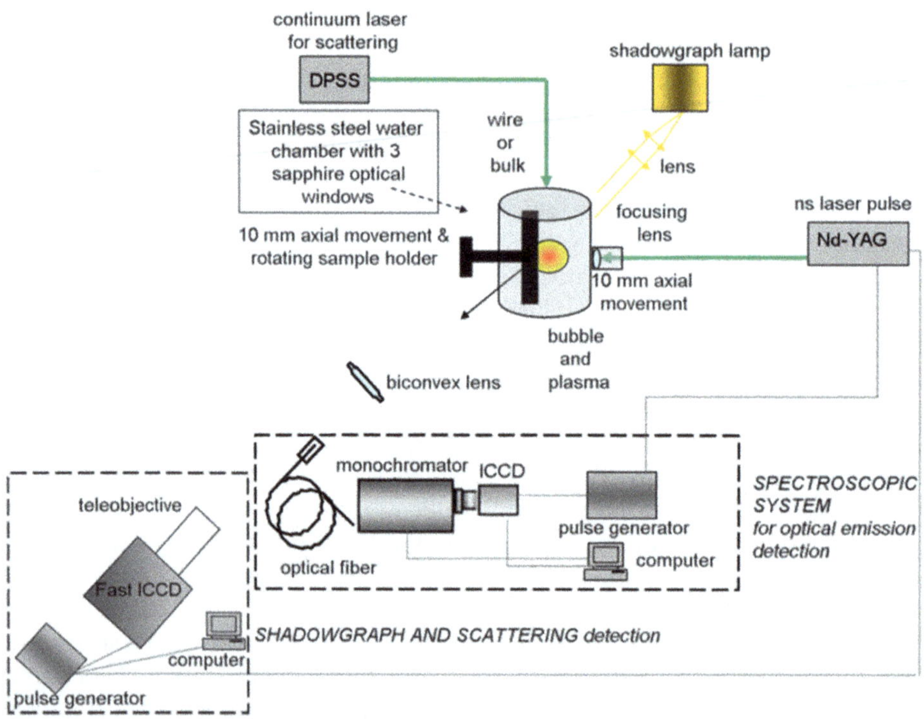

Figure 8.9. Experimental set-up to monitor PLAL nanoparticle generation using optical emission spectroscopy, shadowgraph, and laser scattering. Reprinted from [96] with permission from Royal Society of Chemistry.

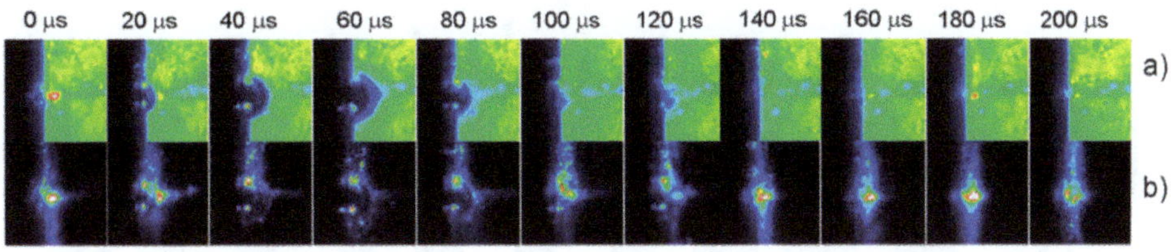

Figure 8.10. Temporally resolved (a) shadowgraph, and (b) laser scattering images of a PLAL bubble on a Ti bulk target in water. Reprinted from [96] with permission from Royal Society of Chemistry.

and laser scattering, images of the process for a titanium target are presented in figure 8.10 [96].

The bubble reaches maximum expansion around 60 μs, and it can be seen that after collapse there is a rebound bubble. The laser scattering signal indicates the produced particles lie close to the target surface. This is undesirable; as discussed particles may deposit onto the surface of the target. It is more desirable that the produced nanoparticles are ejected into the solution. Post-PLAL scanning electron microscope (SEM) examination of the target showed the presence of deposited nanoparticles on the surface. Using the bubble radii determined from the shadowgraph images, the authors estimate the pressure and temperature in the cavitation bubble at each observation time using the van der Waals model [96]:

$$P(t) = \left(P_\infty + \frac{2\sigma}{R}\right)\left(\frac{R_\infty{}^3 - h^3}{R^3 - h^3}\right)^\gamma$$

$$T(t) = T_\infty\left(\frac{R_\infty{}^3 - h^3}{R^3 - h^3}\right)^{\gamma-1}$$

where P_∞ and T_∞ are the pressure and temperature of the liquid, R is the bubble radius, R_∞ is the radius at which the pressure in the bubble corresponds to the liquid pressure, σ is the surface tension of the liquid, $h = R_\infty/9.174$, and $\gamma = 1.22$. From this, they estimated that the pressure and temperature in the bubble exceed 1000 K and 10^7 Pa as the bubble collapses. The values they obtain are limited by the time interval of the shadowgraphs compared to the rapid changes of the bubble. Theoretical modelling of laser-induced cavitation bubble collapse by Akhatov *et al* has shown internal temperature values of 5000–10 000 K, and pressures of thousands of MPa [100]. These high values may be responsible for the deposition of nanoparticles onto the target [96]. Ablation of wire targets was shown to give ejection of the nanoparticles into the solution, unlike the bulk targets, indicating the influence that target shape can have on nanoparticle generation output. For a wire target, the cavitation bubble was found to detach from the surface as it collapses, ejecting the particles. Shadowgraph images for wire targets of increasing thickness and a bulk target can be seen in figure 8.11 [96]. The authors suggest this may affect the formation of the colloid, meriting further investigation, but also improves the output/scalability. The process parameters were also related to the results, indicating that increased repetition rate improves yield. However, if pulses are delivered before the cavitation bubble created by the previous pulse has collapsed, the target will be

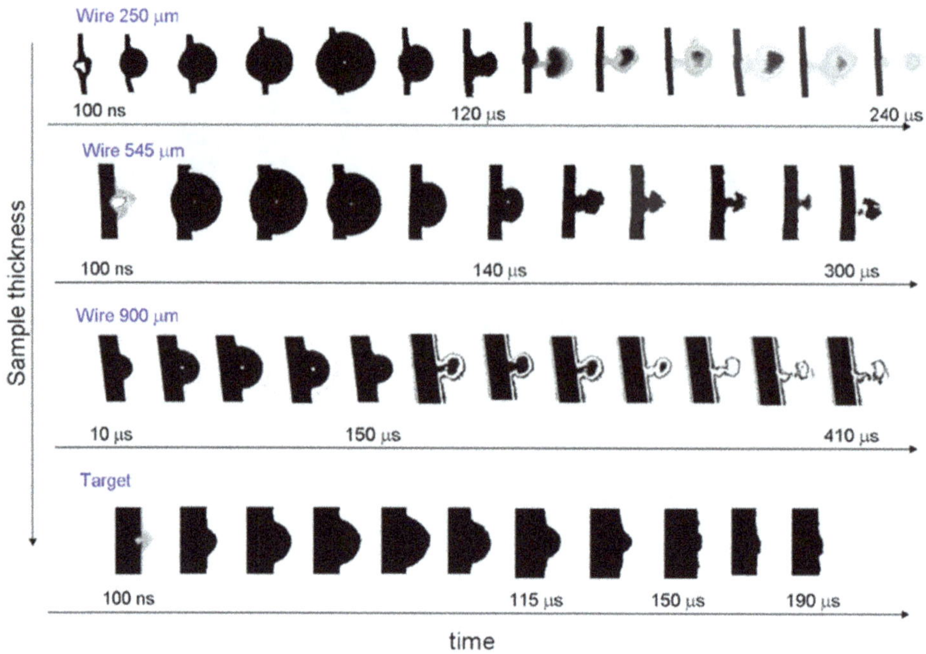

Figure 8.11. Time-resolved shadowgraph images of PLAL cavitation bubbles for different targets. Reprinted from [96] with permission from Royal Society of Chemistry.

shielded from the laser energy by the bubble. The authors estimate, based on the durations determined by shadowgraph, that repetition rates higher than 7 kHz will lead to bubble shielding, and that repetition rates higher than 3 kHz will lead to absorption and scattering by produced nanoparticles, limiting the ablation efficiency.

The majority of PLAL is static, or batch, production: ablation of a solid target in a container of liquid. Nanoparticles are produced in a batch, and then the solution must be removed from the container and replaced with fresh solvent. As discussed, the cavitation bubble and ejected nanomaterials may act to shield the ablation target at higher repetition rates [96]. Furthermore, as production continues, the liquid will begin to saturate with nanoparticles, reducing the ablation efficiency. This static approach has the advantage of simplicity, but limits scalability and output. The low productivity for PLAL has been noted as the bottleneck for the method, inhibiting its uptake in industry [111]. This has driven the development of batch and continuous or semi-continuous methods, where the target is immersed in a flow of solvent [112, 113].

Barcikowski *et al* report on a flow based PLAL method to generate silver nanoparticles [113]. In this method, a magnetic stirrer is utilised to produce flow over the target in a bypass flow cell. While the method is still batch based, the flow of solvent will move nanoparticles, vapour bubbles, and heated liquid away from the ablation site and act to cool the target. A commercial picosecond laser (1064 nm, 10 ps) was focused on a solid silver target in water in the flow cell, scanning the spot linearly. The experiment was carried out with the same processing parameters for static conditions, for comparison. The authors note that the production rate and nanoparticle size distribution for the static process has poor reproducibility. The authors observed a wavy and inconsistent appearance in the ablation lines on the target used under static conditions, and the effective spot size on the surface was found to be larger due to scattering of the laser light. The production rate was increased by 380% for flow conditions compared with static conditions, with picosecond laser ablation achieving a maximum productivity of 8.6 μg s^{-1}.

Messina *et al* report the production of silver nanoparticles with a continuously fed wire target under flow conditions [112]. An Nd:YAG laser (1064 nm, 10 ns) was reduced to a collimated 3 mm using telescope optics and passed through an ablation chamber where a silver wire target (of 125–1000 μm diameter) is continuously fed by a motor, shown in figure 8.12 [112]. The water in the chamber is pumped parallel to the direction of the wire feeding, to avoid bending of the wire by the water pressure, at a flow rate of 9 ml s^{-1}. As discussed with Barcikowski *et al* [113], the geometry of the target was found to have an effect on ablation efficiency, with wire targets giving higher ablation efficiency than bulk targets due to the release of the cavitation bubble from the surface. There are three main factors related to the diameter of the wire influencing the ablation efficiency: the area irradiated by the laser, and thus the fluence; the heat dispersion, which for cylinders decreases with decreasing diameter and will also be affected by the thermal diffusivity of the material; and that for large enough diameters the cavitation bubble behaviour will approach that of a bulk target, with the bubble remaining on the surface as it collapses. In all cases, the wire targets had higher ablation efficiency than a bulk target, and the 750 μm diameter wire was found to give the best efficiency, increased by a factor of 15 compared to a bulk target. Productivity also increased linearly with repetition rates tested, from 1 to 100 Hz, as the interval between pulses was longer than the lifetime of the cavitation bubble (~250 μs). The linear increase should continue, until the interval impinges on the duration of bubble, with maximum repetition rates of about 2 kHz.

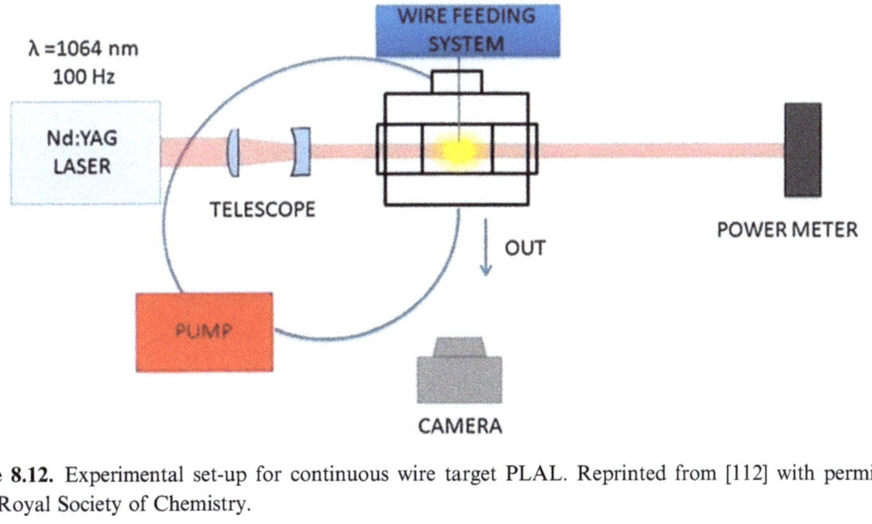

Figure 8.12. Experimental set-up for continuous wire target PLAL. Reprinted from [112] with permission from Royal Society of Chemistry.

The authors estimate that for higher repetition rates a theoretical maximal productivity of >25 g h^{-1} should be achievable with a 750 μm diameter wire target. Fluence (0.4–1.5 J cm^{-2}) was also found to have a linear relationship with the productivity. The silver nanoparticles produced had sizes ranging 15–20 nm, with no difference between those produced from wire and bulk targets, indicating that the temperature and pressure in the cavitation bubble were not affected by the geometry [96, 112].

Streubel *et al* describe a high-speed PLAL process for continuous metal nanoparticle generation [111]. A picosecond laser (1030 nm, 3 ps, 10 MHz, 500 W) combined with a polygon scanner was used to scan the laser spot at speeds of up to 500 m s^{-1} over metal (platinum, silver, gold, aluminium, copper, and titanium) targets under flow conditions. With high scanning speeds, a subsequent pulse can be spatially removed from its prior pulse, allowing higher repetition rates to be used without the issue of cavitation bubble shielding. The authors acknowledge the productivity bottleneck facing PLAL, and point out that studies which report ablation rates $\geqslant$ 1 g h^{-1} often extrapolate from rates measured over seconds or minutes [114, 115]. Schwenke *et al* investigated the influence of processing time on nanoparticle generation [116]. In this work they studied ablation rates by picosecond PLAL for times of up to 1 h. They found evidence that ablated mass did not increase linearly with ablation time, attributed to the absorption or scattering of laser light by the produced nanoparticles, indicating that extrapolating from seconds or minutes of ablation may not be valid. In the work by Streubel *et al*, bypassing the cavitation bubble and nanoparticle shielding using an advanced beam scanner and flow conditions presents an opportunity to achieve high productivity over long time frames (see figure 8.13) [111]. An ultrafast laser beam is directed by two scanning systems, a polygon scanner for fast, vertical translation,

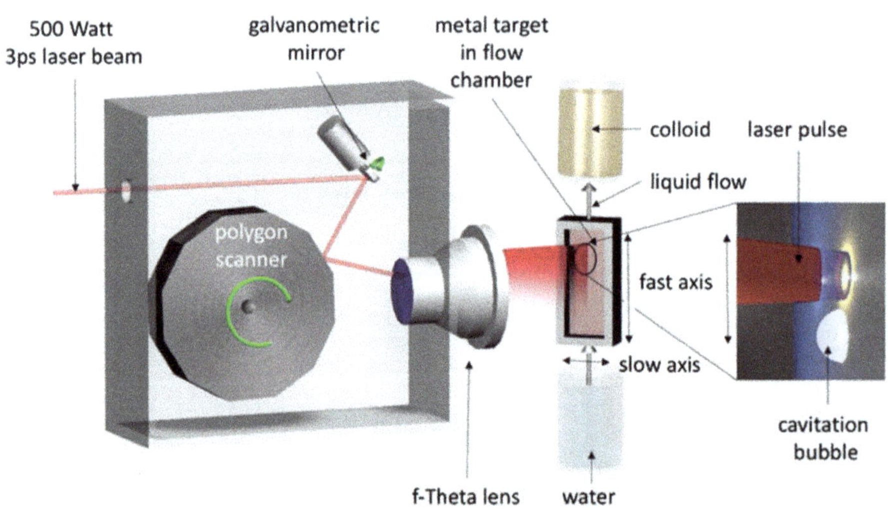

Figure 8.13. Ultrafast polygon scanner PLAL process for continuous metal nanoparticle generation, with a magnified insert showing the spatial bypassing of cavitation bubble shielding. (Reproduced from [111].)

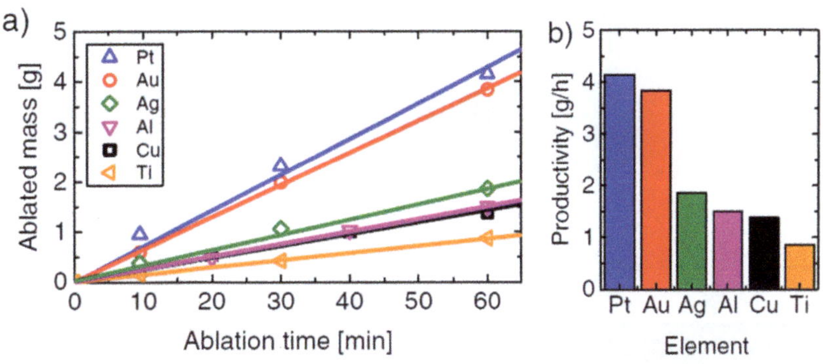

Figure 8.14. Productivities for fast scan PLAL (a) ablated mass as a function of time for six metal targets, (b) the productivity in g h^{-1} for the different targets. (Reproduced from [111].)

and a galvanometric mirror for slower, horizontal translation. Water is pumped through the ablation chamber, with the colloid being collected in a different vessel, allowing for semi-continuous production. The authors carried out optimisation experiments with 10 min durations to optimise the process parameters, then used the optimised parameters (500 W, 10.1 MHz, 484 m s^{-1}) for 1 h of continuous production for six different metal targets (platinum, gold, silver, aluminium, copper, and titanium). The resulting production rates are presented in figure 8.14 [111]. The highest productivity achieved was 4.1 g h^{-1}, for the platinum target.

Another factor affecting the yield for PLAL is the thickness of the liquid layer covering the target [117–119]. As light travels through a liquid medium some of the light will be absorbed, attenuating the beam, with thicker layers of water leading to a greater portion of the laser energy being absorbed. However, if the liquid layer is too thin it may insufficiently confine the plasma plume, even allowing the plume to expand beyond the surface of the liquid [119]. Jiang *et al* investigated the yield for PLAL generation of germanium nanoparticles in deionised water using a 532 nm wavelength Nd:YAG laser [118]. They found the optimum water layer thickness for these conditions to be 1.2 mm.

Kohsakowski *et al* report on a continuous method for nanoparticle generation using wire targets in a liquid jet [102]. Previous works have shown the increased effectiveness and productivity of wire targets [112, 113]. Kohsakowski *et al* present an experimental method for PLAL nanoparticle generation in which a wire target is continuously fed, as in Messina *et al* [112], under a liquid jet. This method combines continuous liquid flow, with a thin liquid layer, a wire target geometry, and continuous feeding of the target. The experimental set-up is shown in figure 8.15 [102]. Wires of different materials and diameters could be fed through the jet nozzle, with interchangeable nozzles of 1–4 mm diameter to allow varying of the liquid layer thickness. Water flow was controlled by a pump to give a flow rate of up to 8 ml s^{-1}, and the layer thickness was measured by imaging with a camera system. A nanosecond Nd:YAG laser (1064 nm, 0.1–15 kHz, 2–8 mJ per pulse) was used for the ablation. A power meter behind the jet was used to ensure good illumination of the wire target by minimising the power detected.

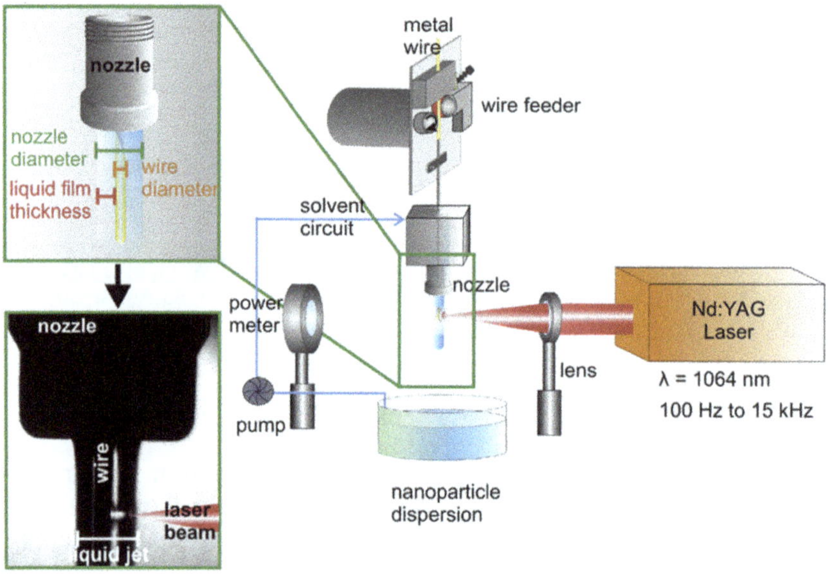

Figure 8.15. Experimental set-up for continuous PLAL using a continuously fed wire target and water jet. Reprinted from [102], copyright (2017) with permission from Elsevier.

The authors determined the process was successful as a continuous and stable nanoparticle production process, with consistent particle sizes. High productivities were obtained, over a 1 h process time: 220, 410, and 550 mg h^{-1} for gold, platinum, and silver, respectively. The values are lower than those obtained by Streubel *et al* [111]. The power specific productivity, in terms of the ablated mass per unit of laser power, was comparable for the two works; ranging 6.77 to 16.92 mg h^{-1} W for Kohsakowski *et al*, and 7.6 to 16.20 mg h^{-1} W for Streubel *et al* [102, 111]. The highest productivities were found with a liquid layer thickness of 500 μm.

8.5 Conclusions

In this chapter, we presented a review of the applications of nanoparticles, a general overview of the generation methods, and a specific review of the PLAL method. PLAL is a useful approach for producing nanomaterials in solution. Control of the processing parameters allows the production of a wide range of sizes and morphologies (nanoparticles, nanorods, etc), and is applicable to a wide range of target materials. The method typically avoids the use of hazardous, environmentally unfriendly chemicals, and produces ligand-free nanoparticles, which are necessary for many applications.

The main limitation for PLAL has been achieving high, commercially viable, production rates. A number of cutting-edge PLAL approaches reviewed here, such as stirred batch, dynamic flow based, high-speed polygon scanning, and wire in liquid jet production have led to success providing greatly improved production rates, which provide the much-needed scalability, enabling the implementation of PLAL as a more commercially viable nanoparticle production method.

Acknowledgements

This research is supported by a research grant from Science Foundation Ireland (SFI) under Grant Number 16/RC/3872 and 19/US-C2C/3579 and is co-funded under the European Regional Development Fund.

References

[1] Navrotsky A 2001 Thermochemistry of nanomaterials *Rev. Mineral Geochem.* **44** 73–103

[2] Balasubramanian K 2010 Challenges in the use of 1D nanostructures for on-chip biosensing and diagnostics: a review *Biosens. Bioelectron.* **26** 1195–204

[3] Tonezzer M, Dang T T L, Bazzanella N, Nguyen V H and Iannotta S 2015 Comparative gas-sensing performance of 1D and 2D ZnO nanostructures *Sens. Actuators, B Chem.* **220** 1152–60

[4] Liu F K 2009 Analysis and applications of nanoparticles in the separation sciences: a case of gold nanoparticles *J. Chromatogr.* A **1216** 9034–47

[5] Park J B, Lee K H, Jeon Y J, Lim S H and Lee S M 2014 Si/C composite lithium-ion battery anodes synthesized using silicon nanoparticles from porous silicon *Electrochim. Acta* **133** 73–81

[6] Yu Y H, Ma C C M, Teng C C, Huang Y L, Lee S H and Wang I *et al* 2012 Electrical, morphological, and electromagnetic interference shielding properties of silver nanowires and nanoparticles conductive composites *Mater. Chem. Phys.* **136** 334–40

[7] Bagga K, McCann R, Wang M, Stalcup a, Vázquez M and Brabazon D 2014 Laser assisted synthesis of carbon nanoparticles with controlled viscosities for printing applications *J. Colloid Interface Sci.*

[8] Haidary S M, Córcoles E P and Ali N K 2012 Nanoporous silicon as drug delivery systems for cancer therapies *J. Nanomater* **2012** 830503

[9] Weber J, Singhal R, Zekri S and Kumar A 2008 One-dimensional nanostructures: fabrication, characterisation and applications *Int. Mater. Rev.* **53** 235–55

[10] Ahmad M, Kiely J, Luxton R, Jabeen M and Khalid M 2016 Facile aqueous growth of 150 nm ZnO nanowires for energy harvester: enhanced output voltage using Pt sputtered electrode *Sens. Bio-Sensing Res.* **7** 141–5

[11] Ahmad M, Iqbal M A, Kiely J, Luxton R and Jabeen M 2017 Enhanced output voltage generation via ZnO nanowires (50 nm): effect of diameter thinning on voltage enhancement *J. Phys. Chem. Solids* **104** 281–5

[12] Sahay S and Kumar M J 2017 Diameter dependence of leakage current in nanowire junctionless field effect transistors *IEEE Trans. Electron Dev.* **64** 1330

[13] Dreaden E C, Alkilany A M, Huang X, Murphy C J and El-Sayed M A 2012 The golden age: gold nanoparticles for biomedicine *Chem. Soc. Rev.* **41** 2740–79

[14] Barcikowski S and Compagnini G 2013 Advanced nanoparticle generation and excitation by lasers in liquids *Phys. Chem. Chem. Phys.* **15** 3022–6

[15] Li Y, Wang Z, Sun L, Liu L, Xu C and Kuang H 2019 Nanoparticle-based sensors for food contaminants *Trends Anal. Chem.* **113** 74–83

[16] El-Ansary A and Faddah L M 2010 Nanoparticles as biochemical sensors *Nanotechnol. Sci. Appl.* **3** 65–76

[17] Sharma R, Ragavan K V, Thakur M S and Raghavarao K S M S 2015 Recent advances in nanoparticle based aptasensors for food contaminants *Biosens. Bioelectron.* **74** 612–27

[18] Deng J, Yu P, Wang Y, Yang L and Mao L 2014 Visualization and quantification of neurochemicals with gold nanoparticles: opportunities and challenges *Adv. Mater.* **26** 6933–43

[19] Bigdeli A, Ghasemi F, Golmohammadi H, Abbasi-Moayed S, Nejad M A F and Fahimi-Kashani N *et al* 2017 Nanoparticle-based optical sensor arrays *Nanoscale* **9** 16546–63

[20] Liu D, Wang Z and Jiang X 2011 Gold nanoparticles for the colorimetric and fluorescent detection of ions and small organic molecules *Nanoscale* **3** 1421–33

[21] Chen X Y, Ha W and Shi Y P 2019 Sensitive colorimetric detection of melamine in processed raw milk using asymmetrically PEGylated gold nanoparticles *Talanta* **194** 475–84

[22] Lam C W, Lan L, Che X, Tam S, Wong S S Y and Chen Y *et al* 2009 Diagnosis and spectrum of melamine-related renal disease: plausible mechanism of stone formation in humans *Clin. Chim. Acta* **402** 150–5

[23] Ramezani M, Mohammad Danesh N, Lavaee P, Abnous K and Mohammad Taghdisi S 2015 A novel colorimetric triple-helix molecular switch aptasensor for ultrasensitive detection of tetracycline *Biosens. Bioelectron.* **70** 181–7

[24] Ruedas-Rama M J, Walters J D, Orte A and Hall E A H 2012 Fluorescent nanoparticles for intracellular sensing: a review *Anal. Chim. Acta* **751** 1–23

[25] Xu J, Li Y, Wang L, Huang Y, Liu D and Sun R *et al* 2015 A facile aptamer-based sensing strategy for dopamine through the fluorescence resonance energy transfer between rhodamine B and gold nanoparticles *Dye Pigment* **123** 55–63

[26] Konduru T, Rains G C and Li C 2015 A customized metal oxide semiconductor-based gas sensor array for onion quality evaluation: system development and characterization *Sensors (Basel)* **15** 1252–73

[27] Cigić I K and Prosen H 2009 An overview of conventional and emerging analytical methods for the determination of mycotoxins *Int. J. Mol. Sci.* **10** 62–115

[28] Vargas-Bernal R, Rodrguez-Miranda E and Herrera-Prez G 2012 Evolution and expectations of enzymatic biosensors for pesticides *Advances in Chemical and Botanical Pesticides* ed R P Soundararajan (Rijeka: IntechOpen)

[29] Yusoff N, Pandikumar A, Ramaraj R, Lim H N and Huang N M 2015 Gold nanoparticle based optical and electrochemical sensing of dopamine *Microchim. Acta* **182** 2091–114

[30] Raj C R, Okajima T and Ohsaka T 2003 Gold nanoparticle arrays for the voltammetric sensing of dopamine *J. Electroanal. Chem.* **543** 127–33

[31] Cruz S M F, Rocha L A and Viana J C 2018 Printing technologies on flexible substrates for printed electronics *Flexible Electronics* ed S Rackauskas (Rijeka: IntechOpen)

[32] Nayak L, Mohanty S, Nayak S K and Ramadoss A 2019 A review on inkjet printing of nanoparticle inks for flexible electronics *J. Mater. Chem.* C **7** 8771–95

[33] Gao Y, Liu R, Wang X, Liu J and Fang Q 2016 Flexible RFID tag inductor printed by liquid metal ink printer and its characterization *J. Electron. Packag.* **138** 031007

[34] Kopyt P, Salski B, Olszewska M, Janczak D, Sloma M and Kurkus T *et al* 2015 Graphene-based dipole antenna for a UHF RFID tag *Proc. 2015 IEEE MTT-S Int Microw Symp.* **1** 1–3

[35] Leng T, Huang X, Chang K, Chen J, Abdalla M A and Hu Z 2016 Graphene nanoflakes printed flexible meandered-line dipole antenna on paper substrate for low-cost RFID and sensing applications *IEEE Antennas Wirel. Propag. Lett.* **15** 1565–8

[36] Matyas J, Munster L, Olejnik R, Vlcek K, Slobodian P and Krcmar P *et al* 2016 Antenna of silver nanoparticles mounted on a flexible polymer substrate constructed using inkjet print technology *Jpn. J. Appl. Phys.* **55** 02BB13

[37] Jeong S, Song H C, Lee W W, Lee S S, Choi Y and Son W *et al* 2011 Stable aqueous based Cu nanoparticle ink for printing well-defined highly conductive features on a plastic substrate *Langmuir* **27** 3144–9

[38] Duval J F and Herr H M 2016 FlexSEA: flexible, scalable electronics architecture for wearable robotic applications *Proc. IEEE RAS EMBS Int. Conf. Biomed Robot Biomechatronics* 1236–41

[39] Cheng M Y, Lin C L and Yang Y J 2010 Tactile and shear stress sensing array using capacitive mechanisms with floating electrodes *Proc. IEEE Int. Conf. Micro Electro Mech. Syst.* **2** 228–31

[40] Chiang C C, Lin CC K and Ju M S 2007 An implantable capacitive pressure sensor for biomedical applications *Sens. Actuators, A Phys.* **134** 382–8

[41] Kim K, Kim B and Lee C H 2019 Printing flexible and hybrid electronics for human skin and eye-interfaced health monitoring systems *Adv. Mater.* **1902051** 1–22

[42] Komoda N, Nogi M, Suganuma K, Kohno K, Akiyama Y and Otsuka K 2012 Printed silver nanowire antennas with low signal loss at high-frequency radio *Nanoscale* **4** 3148

[43] He H, Sydänheimo L, Virkki J and Ukkonen L 2016 Experimental study on inkjet-printed passive UHF RFID tags on versatile paper-based substrates *Int. J Antennas Propag.* **2016** 9265159

[44] Sydanheimo L, Virkki J, Ukkonen L and Ren Y 2014 Optimisation of manufacturing parameters for inkjet-printed and photonically sintered metallic nanoparticle UHF RFID tags *Electron. Lett.* **50** 1504–5

[45] Chen B, Jiang Y, Tang X, Pan Y and Hu S 2017 Fully packaged carbon nanotube supercapacitors by direct ink writing on flexible substrates *ACS Appl. Mater. Interfaces* **9** 28433–40

[46] Rajan K, Roppolo I, Chiappone A, Bocchini S, Perrone D and Chiolerio A 2016 Nanotechnology, science and applications dovepress silver nanoparticle ink technology: state of the art *Nanotechnol. Sci. Appl.* **9** 1–13

[47] Balliu E, Andersson H, Engholm M, Öhlund T, Nilsson H E and Olin H 2018 Selective laser sintering of inkjet-printed silver nanoparticle inks on paper substrates to achieve highly conductive patterns *Sci. Rep.* **8** 2–10

[48] Taufik S, Barfidokht A, Alam M T, Jiang C, Parker S G and Gooding J J 2016 An antifouling electrode based on electrode–organic layer–nanoparticle constructs: electrodeposited organic layers versus self-assembled monolayers *J. Electroanal. Chem.* **779** 229–35

[49] Nguyen A, Zou L and Priest C 2014 Evaluating the antifouling effects of silver nanoparticles regenerated by TiO_2 on forward osmosis membrane *J. Membr. Sci.* **454** 264–71

[50] Kirschner C M and Brennan A B 2012 Bio-inspired antifouling strategies *Annu. Rev. Mater. Res.* **42** 211–29

[51] Yu B, Leung K M, Guo Q, Lau W M and Yang J 2011 Synthesis of Ag–TiO_2 composite nano thin film for antimicrobial application *Nanotechnology* **22** 115603

[52] Schacht V J, Neumann L V, Sandhi S K, Chen L, Henning T and Klar P J *et al* 2013 Effects of silver nanoparticles on microbial growth dynamics *J. Appl. Microbiol.* **114** 25–35

[53] Yang H L, Lin J C T and Huang C 2009 Application of nanosilver surface modification to RO membrane and spacer for mitigating biofouling in seawater desalination *Water Res.* **43** 3777–86

[54] Sondi I and Salopek-Sondi B 2004 Silver nanoparticles as antimicrobial agent: a case study on *E. coli* as a model for Gram-negative bacteria *J. Colloid Interface Sci.* **275** 177–82

[55] Marambio-Jones C and Hoek E M V 2010 A review of the antibacterial effects of silver nanomaterials and potential implications for human health and the environment *J. Nanoparticle Res.* **12** 1531–51

[56] AshaRani P V, Mun G L K, Hande M P and Valiyaveettil S 2009 Cytotoxicity and genotoxicity of silver nanomaterials *ACS Nano* **3** 279–90

[57] Lok C N, Ho C M, Chen R, He Q Y, Yu W Y and Sun H *et al* 2007 Silver nanoparticles: partial oxidation and antibacterial activities *J. Biol. Inorg. Chem.* **12** 527–34

[58] Pi J K, Yang H C, Wan L S, Wu J and Xu Z K 2016 Polypropylene microfiltration membranes modified with TiO_2 nanoparticles for surface wettability and antifouling property *J. Membr. Sci.* **500** 8–15

[59] de la Torre C and Ceña V 2018 The delivery challenge in neurodegenerative disorders: the nanoparticles role in Alzheimer's disease therapeutics and diagnostics *Pharmaceutics* **17** 190

[60] Wen M M, El-Salamouni N S, El-Refaie W M, Hazzah H A, Ali M M and Tosi G *et al* 2017 Nanotechnology-based drug delivery systems for Alzheimer's disease management: technical, industrial, and clinical challenges *J. Control. Release* **245** 95–107

[61] Cunha S, Amaral M, Lobo J and Silva A 2016 Therapeutic strategies for Alzheimer's and Parkinson's diseases by means of drug delivery systems *Curr. Med. Chem.* **23** 3618–31

[62] Xiao Y, Shi K, Qu Y, Chu B and Qian Z 2019 Engineering nanoparticles for targeted delivery of nucleic acid therapeutics in tumor *Mol. Ther. Methods Clin. Dev.* **12** 1–18

[63] Singh P, Pandit S, Mokkapati V R S S, Garg A, Ravikumar V and Mijakovic I 2018 Gold nanoparticles in diagnostics and therapeutics for human cancer *Int. J. Mol. Sci.* **19** 1979

[64] Hui Y, Yi X, Hou F, Wibowo D, Zhang F and Zhao D *et al* 2019 Role of nanoparticle mechanical properties in cancer drug delivery *ACS Nano* **13** 7410–24

[65] Lewis D R, Kamisoglu K, York A W and Moghe P V 2011 Polymer-based therapeutics: nanoassemblies and nanoparticles for management of atherosclerosis *Interdiscip. Rev. Nanomed. Nanobiotechnol.* **3** 400–20

[66] Bharali D J, Klejbor I, Stachowiak E K, Dutta P, Roy I and Kaur N *et al* 2005 Organically modified silica nanoparticles: a nonviral vector for *in vivo* gene delivery and expression in the brain *Proc. Natl. Acad. Sci. U.S.A.* **102** 11539–44

[67] Huang X and El-Sayed M A 2011 Plasmonic photo-thermal therapy (PPTT) *Alexandria J. Med.* **47** 1–9

[68] Ali M R K, Wu Y and El-Sayed M A 2019 Gold-nanoparticle-assisted plasmonic photothermal therapy advances toward clinical application *J. Phys. Chem.* C **123** 15375–93

[69] Napier M E and DeSimone J M 2007 Nanoparticle drug delivery platform *Polym. Rev.* **47** 321–7

[70] Liu M, Song W and Huang L 2019 Drug delivery systems targeting tumor-associated fibroblasts for cancer immunotherapy *Cancer Lett.* **448** 31–9

[71] Bahrami B, Hojjat-Farsangi M, Mohammadi H, Anvari E, Ghalamfarsa G and Yousefi M *et al* 2017 Nanoparticles and targeted drug delivery in cancer therapy *Immunol. Lett.* **190** 64–83

[72] Kong F Y, Zhang J W, Li R F, Wang Z X, Wang W J and Wang W 2017 Unique roles of gold nanoparticles in drug delivery, targeting and imaging applications *Molecules* **22** 1445

[73] Ajnai G, Chiu A, Kan T, Cheng C C, Tsai T H and Chang J 2014 Trends of gold nanoparticle-based drug delivery system in cancer therapy *J. Exp. Clin. Med.* **6** 172–8

[74] Chen Y H, Tsai C Y, Huang P Y, Chang M Y, Cheng P C and Chou C H *et al* 2007 Methotrexate conjugated to gold nanoparticles inhibits tumor growth in a syngeneic lung tumor model *Mol. Pharm.* **4** 713–22

[75] Wechsler M E, Vela Ramirez J E and Peppas N A 2019 110th Anniversary: nanoparticle mediated drug delivery for the treatment of Alzheimer's disease: crossing the blood–brain barrier *Ind. Eng. Chem. Res.* **58** 15079–87

[76] Han P, Martens W, Waclawik E R, Sarina S and Zhu H 2018 Metal nanoparticle photocatalysts: synthesis, characterization, and application *Part. Part. Syst. Charact.* **35** 1–16

[77] Wiley B, Sun Y, Mayers B and Xia Y 2005 Shape-controlled synthesis of metal nanostructures: the case of silver *Chem. A Eur. J.* **11** 454–63

[78] Iravani S, Korbekandi H, Mirmohammadi S V and Zolfaghari B 2014 Synthesis of silver nanoparticles: chemical, physical and biological methods *Res. Pharm. Sci.* **9** 385–406

[79] Oliveira M M, Ugarte D, Zanchet D and Zarbin A J G 2005 Influence of synthetic parameters on the size, structure, and stability of dodecanethiol-stabilized silver nanoparticles *J. Colloid Interface Sci.* **292** 429–35

[80] Zhao P, Li N and Astruc D 2013 State of the art in gold nanoparticle synthesis *Coord. Chem. Rev.* **257** 638–65

[81] Turkevich J, Stevenson P C and Hillier J 1951 A study of the nucleation and growth processes in the synthesis of colloidal gold *Discuss. Faraday Soc.* **11** 55–75

[82] Frens G 1973 Controlled nucleation for the regulation of the particle size in monodisperse gold suspensions *Nat. Phys. Sci.* **241** 20–2

[83] Brown K R, Fox A P and Natan M J 1996 Morphology-dependent electrochemistry of cytochrome c at Au colloid-modified SnO_2 electrodes *J. Am. Chem. Soc.* **118** 1154–7

[84] Roy A, Bulut O, Some S, Mandal A K and Yilmaz M D 2019 Green synthesis of silver nanoparticles: biomolecule–nanoparticle organizations targeting antimicrobial activity *RSC Adv.* **9** 2673–702

[85] Mittal A K, Chisti Y and Banerjee U C 2013 Synthesis of metallic nanoparticles using plant extracts *Biotechnol. Adv.* **31** 346–56

[86] Li X, Xu H, Chen Z S and Chen G 2011 Biosynthesis of nanoparticles by microorganisms and their applications *J. Nanomater* **2011** 270974

[87] Dwivedi A D and Gopal K 2010 Biosynthesis of silver and gold nanoparticles using *Chenopodium album* leaf extract *Colloids Surf. A Physicochem. Eng. Asp.* **369** 27–33

[88] Magnusson M H, Deppert K, Malm J O, Bovin J O and Samuelson L 1999 Gold nanoparticles: production, reshaping, and thermal charging *J. Nanoparticle Res.* **1** 243–51

[89] Lo C H, Tsung T T and Lin H M 2007 Preparation of silver nanofluid by the submerged arc nanoparticle synthesis system (SANSS) *J. Alloys Compd.* **434–435** 659–62

[90] Lung J K, Huang J C, Tien D C, Liao C Y, Tseng K H and Tsung T T *et al* 2007 Preparation of gold nanoparticles by arc discharge in water *J. Alloys Compd.* **434–435** 655–8

[91] Wang L L and Jiang J S 2007 Preparation of α-Fe_2O_3 nanoparticles by high-energy ball milling *Phys. B Condens. Matter* **390** 23–7

[92] Pimpang P, Sutham W, Mangkorntong N, Mangkorntong P and Choopun S 2008 Effect of stabilizer on preparation of silver and gold nanoparticle using grinding method *Chiang Mai J. Sci.* **35** 250–7

[93] Arbain R, Othman M and Palaniandy S 2011 Preparation of iron oxide nanoparticles by mechanical milling *Miner. Eng.* **24** 1–9

[94] Knieke C, Steinborn C, Romeis S, Peukert W, Breitung-Faes S and Kwade A 2010 Nanoparticle production with stirred-media mills: opportunities and limits *Chem. Eng. Technol.* **33** 1401–11

[95] Baláž M, Balážová Ľ, Daneu N, Dutková E, Balážová M and Bujňáková Z *et al* 2017 Plant-mediated synthesis of silver nanoparticles and their stabilization by wet stirred media milling *Nanoscale Res. Lett.* **12** 83

[96] De Giacomo A, Dell'Aglio M, Santagata A, Gaudiuso R, De Pascale O and Wagener P *et al* 2013 Cavitation dynamics of laser ablation of bulk and wire-shaped metals in water during nanoparticles production *Phys. Chem. Chem. Phys.* **15** 3083–92

[97] Dell'Aglio M, Gaudiuso R, De Pascale O and De Giacomo A 2015 Mechanisms and processes of pulsed laser ablation in liquids during nanoparticle production *Appl. Surf. Sci.* **348** 4–9

[98] Lauterborn W and Kurz T 2010 Physics of bubble oscillations *Reports Prog. Phys.* **73** 106501

[99] Petkovšek R and Gregorčič P 2007 A laser probe measurement of cavitation bubble dynamics improved by shock wave detection and compared to shadow photography *J. Appl. Phys.* **102** 044909

[100] Akhatov I, Lindau O, Topolnikov A, Mettin R, Vakhitova N and Lauterborn W 2001 Collapse and rebound of a laser-induced cavitation bubble *Phys. Fluids* **13** 2805–19

[101] Zeng H, Du X W, Singh S C, Kulinich S A, Yang S and He J *et al* 2012 Nanomaterials via laser ablation/irradiation in liquid: a review *Adv. Funct. Mater.* **22** 1333–53

[102] Kohsakowski S, Santagata A, Dell'Aglio M, de Giacomo A, Barcikowski S and Wagener P *et al* 2017 High productive and continuous nanoparticle fabrication by laser ablation of a wire-target in a liquid jet *Appl. Surf. Sci.* **403** 487–99

[103] Ibrahimkutty S, Wagener P, Menzel A, Plech A and Barcikowski S 2012 Nanoparticle formation in a cavitation bubble after pulsed laser ablation in liquid studied with high time resolution small angle x-ray scattering *Appl. Phys. Lett.* **101** 103104

[104] Kabashin A V and Meunier M 2003 Synthesis of colloidal nanoparticles during femtosecond laser ablation of gold in water *J. Appl. Phys.* **94** 7941–3

[105] Sylvestre J P, Kabashin A V, Sacher E and Meunier M 2005 Femtosecond laser ablation of gold in water: influence of the laser-produced plasma on the nanoparticle size distribution *Appl. Phys. A Mater. Sci. Process* **80** 753–8

[106] Gamrad L, Rehbock C, Krawinkel J, Tumursukh B, Heisterkamp A and Barcikowski S 2014 Charge balancing of model gold–nanoparticle–peptide conjugates controlled by the peptide's net charge and the ligand to nanoparticle ratio *J. Phys. Chem.* C **118** 10302–13

[107] Marzun G, Nakamura J, Zhang X, Barcikowski S and Wagener P 2015 Size control and supporting of palladium nanoparticles made by laser ablation in saline solution as a facile route to heterogeneous catalysts *Appl. Surf. Sci.* **348** 75–84

[108] Mao S S, Mao X, Greif R and Russo R E 2000 Initiation of an early-stage plasma during picosecond laser ablation of solids *Appl. Phys. Lett.* **77** 2464–6

[109] Plech A, Kotaidis V, Lorenc M and Wulff M 2005 Thermal dynamics in laser excited metal nanoparticles *Chem. Phys. Lett.* **401** 565–9

[110] Shih C Y, Streubel R, Heberle J, Letzel A, Shugaev M V and Wu C *et al* 2018 Two mechanisms of nanoparticle generation in picosecond laser ablation in liquids: the origin of the bimodal size distribution *Nanoscale* **10** 6900–10

[111] Streubel R, Barcikowski S and Gökce B 2016 Continuous multigram nanoparticle synthesis by high-power, high-repetition-rate ultrafast laser ablation in liquids *Opt. Lett.* **41** 1486

[112] Messina G C, Wagener P, Streubel R, De Giacomo A, Santagata A and Compagnini G *et al* 2013 Pulsed laser ablation of a continuously-fed wire in liquid flow for high-yield production of silver nanoparticles *Phys. Chem. Chem. Phys.* **15** 3093–8

[113] Barcikowski S, Meńndez-Manjón A, Chichkov B, Brikas M and Račiukaitis G 2007 Generation of nanoparticle colloids by picosecond and femtosecond laser ablations in liquid flow *Appl. Phys. Lett.* 91

[114] Bärsch N, Jakobi J, Weiler S and Barcikowski S 2009 Pure colloidal metal and ceramic nanoparticles from high-power picosecond laser ablation in water and acetone *Nanotechnology* **20** 44

[115] Sajti C L, Sattari R, Chichkov B and Barcikowski S 2010 Ablation efficiency of α-Al_2O_3 in liquid phase and ambient air by nanosecond laser irradiation *Appl. Phys. A Mater. Sci. Process* **100** 203–6

[116] Schwenke A, Wagener P, Nolte S and Barcikowski S 2011 Influence of processing time on nanoparticle generation during picosecond-pulsed fundamental and second harmonic laser ablation of metals in tetrahydrofuran *Appl. Phys. A Mater. Sci. Process* **104** 77–82

[117] Bärsch N 2009 Improving laser ablation of zirconia by liquid films: multiple influence of liquids on surface machining and nanoparticle generation *J. Laser Micro/Nanoeng.* **4** 66–70

[118] Jiang Y, Liu P, Liang Y, Li H B and Yang G W 2011 Promoting the yield of nanoparticles from laser ablation in liquid *Appl. Phys. A Mater. Sci. Process* **105** 903–7

[119] Nguyen T T P, Tanabe-Yamagishi R and Ito Y 2019 Impact of liquid layer thickness on the dynamics of nano- to sub-microsecond phenomena of nanosecond pulsed laser ablation in liquid *Appl. Surf. Sci.* **470** 250–8

IOP Publishing

Laser Micro- and Nano-Scale Processing
Fundamentals and applications
Ahmed Issa and Dermot Brabazon

Chapter 9

Effect of laser surface treatment on solar cell efficiency

Fatema H Rajab, Ahmad W AlShaer and Tayf T A Sahib

Laser surface treatment is one of the most efficient techniques for the generation of a range of micro-/nano-structures on different material surfaces. This chapter discusses laser modification of solar cell materials surface topographies and the mechanism behind micro- and nano-surface patterning and laser–material interaction. Among the available renewable energy candidates, solar energy is much less site-dependent and it has been considered as an ideal energy source for a large number of applications. The solar cells, their types, manufacturing techniques and their applications are reviewed. Compared to conventional solar cell modification techniques, laser processing is rapid, and it can provide a considerable increase in solar cell efficiency. The use of laser surface treatment over other techniques is critically evaluated in this chapter.

9.1 Introduction

Recently, with the increase in environmental problems such as global warming and damage to the ozone layer due to the use of conventional energy sources such as fossil fuels, the demand for renewable energy resources has been increased. There are several renewable energy resources like wind and nuclear power; however, the source of solar energy has advantages such as sustainability, abundance and being environmentally friendly compared to other resources. Solar cells are photovoltaic devices that convert solar energy into electricity [1]. Their operation is based on splitting the charge carrier at the interface of two opposite materials [2].

There are three main generations of solar cells. The first generation is the conventional solar cells, which are a single junction of polycrystalline, or n-crystalline silicon solar cells. This first solar cell was invented in 1950. The efficiency of this type of solar cell is dependent on the crystalline purity and the maximum efficiency achieved is 20%. This type of solar cell suffers from drawbacks like energy consumption and material

doi:10.1088/978-0-7503-1683-5ch9

purity requirements. Moreover, the cost and manufacturing quality are the main problems limiting its use [3–5]. Thin film photovoltaic cells of mono-junctions or multi-junctions such as multicrystal collar cells, amorphous Si and CdTc solar cells are the second generation of solar cells [3]. Although the toxicity of thin films is considerably high, these cells are easy to manufacture with an efficiency of 10%–15% [6]. Organic solar cells, dye-sensitised solar cells (DSSCs), quantum-dot-sensitised solar cells and perovskite solar cells are the third generation of solar cells [3]. The cost effectiveness and ease of manufacturing with high efficiency (20% for perovskite solar cells and 13% of the DSSCs) can replace the use of other types of solar cells. However, the toxicity of perovskite solar cells is high compared with that of DSSCs. The high toxicity limits the applications of perovskite solar cells compared with DSSCs. For these reasons, this chapter will focus on reviewing the construction, characteristics, manufacturing processes and efficiency enhancement methods of DSSCs.

9.2 Dye-sensitised solar cells

9.2.1 The construction of DSSCs

DSSCs are photo-electrochemical solar cells which were invented to eliminate the drawbacks of conventional solar cells. DSSCs are considered to be non-toxic, cost effective and easy to manufacture and were invented in 1991. Conventional Si solar cells exhibit low efficiency under low light conditions. However, DSSCs are independent of the incident light and perform at high efficiency even under low light conditions. This enables the use of DSSCs in different applications such as intelligent buildings and smart houses [7]. However, to date, the maximum efficiency of DSSCs is 13% [8, 9].

DSSCs consist of a photoanode, electrolyte and photocathode (figure 9.1). The photoanode is the heart of the DSSCs and it is a thin film layer of a semiconductor like TiO_2 and ZnO which is deposited on a conductive substrate like ITO and FTO glass. This thin film works as an adsorbing layer of dye molecules. The dye that can be used in solar cells, which is usually Ru-dye, should strongly adsorb the light in the visible and near-infrared range and should be anchored on the film material. Moreover, it should have an energy level with a good alignment with the band gap of the film material to get a good injection of the excited electrons in the conduction band. Moreover, the dye should have a good thermal and chemical

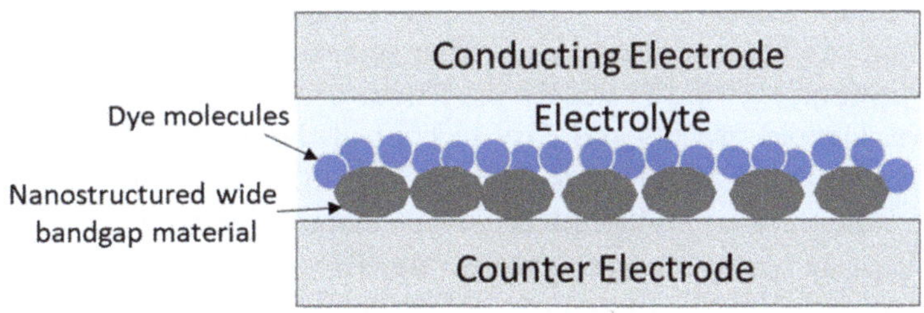

Figure 9.1. The construction of DSSCs.

stability [10, 11]. The electrolyte layer is mediated by the photoanode and cathode electrodes. The cathode electrode is a platinum counter electrode [12].

TiO_2 is widely used as a semiconductor due to its characteristics such as:

1. It is an n-doped semiconductor with a high band gap of 3.2 eV and a large refractive index (2.4–2.5), which promotes efficient transmission of the visible light.
2. It has high chemical stability, affordability and non-toxicity helps in using it in various applications such as photo catalysts, electronic devices, photo electrodes and environmental purifiers [13].
3. Rutile, brookite and anatase are the three main crystalline polymorphous existence of TiO_2 [14].
4. Fast charge carrier injection rate [15].
5. Rapid collection electrons at the anode substrate [15].

Thin film affects the efficiency of the solar cell and the efficiency of solar cells can be increased by [16]:

1. Increasing the surface area of the thin film.
2. Providing a fast electron path way by enhancing the non-crystalline structure.
3. Producing a porous structure by a sintering process.

9.2.2 The fabrication of DSSCs

The most important thing in the fabrication of the DSSCs is the fabrication of the mesoporous film. For fabricating the mesoporous film on a substrate, the film is first deposited on the substrate followed by a sintering process using a furnace with a temperature of 450–500 °C for 30 min or any other sintering processes to achieve sufficient interconnection between nanoparticles forming the mesoporous structure and to ensure a good adhesion between the film and the substrate [17].

There are several methods for depositing mesoporous thin film such as spray pyrolysis, spin coating, thermal oxidation, atomic layer deposition and pulse laser deposition [18]. Among these techniques, doctor blading and sol–gel templating are the two most widely used approaches to prepare thin films. Several techniques are used for the sintering process such as microwave, furnace, near-infrared and laser treatment [19–26].

Before depositing the thin film on the glass substrate, a colloidal paste is prepared by mixing the nanoparticles with organic solvent and organic binder. The deposition of the thin film can be conducted using either screen-printing or sol–gel templating. The screen-printing method is widely used due to it being environmentally clean, cost effective, a rapid process and easy to control the film thickness [27, 28]. This method is widely used when thick film 10–15 μm is required [29]. During this method, the colloidal paste is first placed over the mesh screen. Then, using a rubber squeegee, the paste is spread over the substrate. The sintering process is followed by heating up to 450–500 °C to remove the organic binder from the mesoporous structure [29]. Doctor blading or the tape casting technique is another method that can be used to achieve this film thickness. It is also characterised as a simple method that is environmentally clean and easy to achieve multi-uniform layers [8]. Briefly, the paste is placed on the substrate followed by spreading of the paste using a glass rod. Then, high temperature sintering is used to evaporate the binder and form a mesoporous structure [29].

To remove the organic binder from the paste and to achieve necking between nanoparticles, the sintering process is used. Sintering temperature and time are the two important parameters that affect the solar cell efficiency. It has been reported that for achieving an effective solar cell, sintering using 450–500 °C temperatures for 30 min to 1 h are used.

The sintering process can be carried out using several methods:

1. Conventional thermal method.

   This can be performed using an oven by heating the sample up to 450–500 °C for a period of 30 min. This method results in a good interconnection between nanoparticles and strong adhesion to the substrate [30]. However, this method is not preferable for thermal sensitive materials such as glass in which it results sometimes in cracking or binding the material under high temperature for a long time [31].

2. Alternative methods.

   Aiming to save time and energy and for the possibility of applications for large-scale applications, alternative methods can be used. One of these methods is to manufacture binder-free paste using low temperature sintering [32]. The authors suggested the name of chemical sintering or mechanical compression for this method [33, 34]. This method saved the temperature while the conversion efficiency was lower than the conventional standard method [34]. Another approach is using microwave heating to reduce the sintering time. This method increased the efficiency or achieved the same efficiency as the conventional method. However, it resulted in cracking of the substrate [19]. Pressure plasma jet way is another way that could be used for sintering at 500 °C with a very short time, such as 60 s, which is equivalent to 15 min of oven sintering [35]. Using near-infrared radiation, sintering was achieved for 12.5 s, which was equivalent to 30 min using oven sintering [36]. Laser sintering is an approach that can be used to achieve more precise and highly localised sintering than other sintering processes [37].

9.3 Laser surface treatment

Laser surface treatment of materials or laser material processing is a technique that offers a possibility of enhancing various surface properties such as the surface hardness, strength, coefficient of friction, roughness, corrosion and chemical resistance. A laser generates a beam of high energy in a localised region, which makes it an effective tool for several applications such as welding, cutting, drilling, shaping, cladding, sintering, marking, texturing, etc. Some of these applications, like cutting and drilling, require material removal while others, like sintering and welding, require material melting [38].

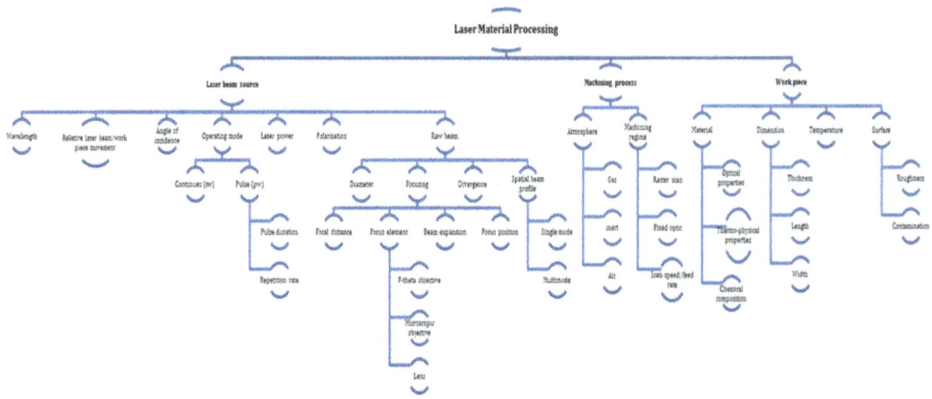

Figure 9.2. Parameters affecting laser material processing.

These applications depend on laser properties such as laser fluence/density, pulse duration, wavelength and material properties such as thermophysical and optical properties (figure 9.2). Therefore, for achieving the required results during laser material processing, efforts should be made to optimise the processing parameters [38].

In laser processing of materials, two different phenomena may take place, namely, photolytic (photothermal) and pyrolytic (photochemical) processes [38]. The photothermal process is based on material removal by heating with or without material melting, whereas the photochemical process is based on material removal by breaking the chemical bonds between the material's molecules [38]. When the laser beam reaches the material surface, a proportion of its energy is reflected while the rest is absorbed. The energy is mainly absorbed by the free and excited electrons, forming a non-equilibrium state. Then thermalisation occurs due to electron–electron or electron–photon interaction. Then, as a result of transferring the energy to the bulk material, the energy is redistributed and the equilibrium state is regained [39, 40]. With the high laser fluence values, the surface vaporises and plasma forms close to the surface in comparison to the surface melting that is observed when lower values of fluence are used [41, 42]. Surface structure and its properties, such as colour, reflectivity, absorptivity, conductivity, corrosion, chemistry and wettability can be altered by the use of the correct processing parameters. For example, figures 9.3 to 9.7 show a range of surface topographies that have been generated on different substrates using different laser parameters and material properties. Figure 9.3 shows a processed stainless steel surface using a picosecond laser beam with different scanning speeds. By changing the scanning speed, the surface colour can be changed and turned into black surfaces using a scanning speed of 1 mm s^{-1}. Figure 9.4 shows the effect of changing the substrate on a micro-/nano-surface structure. It can be seen that scanning the substrates using the same laser parameters generated different surface topography. Figure 9.5 shows the effect of scanning environment on the surface topography using the same substrate. The generated topography on air was a cauliflower-like structure with high roughness while it was a cone-like structure with

500 mm/s 1 mm/s

Figure 9.3. Changing the surface colour by changing the scanning speed.

Zn W SS

Figure 9.4. Nanosecond laser surface treatment of different substrates using a scanning speed of 1 mm s^{-1}, laser power 7 W and pulse duration of 7 ns.

air water

Figure 9.5. Nanosecond laser surface treatment of stainless steel in different environments. A cauliflower-like structure was achieved by processing the surfaces in air while a cone-like structure was achieved by processing the surface in water.

smaller features and lower roughness in water compared with that generated in air. Figure 9.6 shows the effect of the scanning speed on surface topography. It is clear that the surface topography ranged from micro cones covered by sub-micron and nano features using a low scanning speed (10 mm s^{-1}) to a laser induced periodic surface structure (LIPSS) using a relatively high scanning speed (1000 mm s^{-1}) [43].

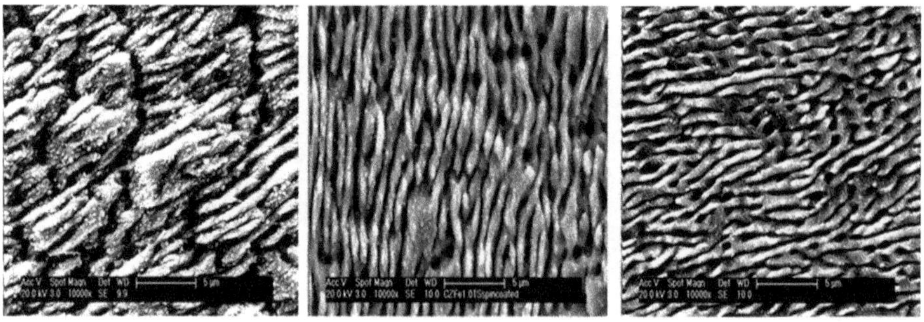

10 mm/s   50 mm/s   1000 mm/s

Figure 9.6. Picosecond laser surface treatment of stainless steel using different scanning speeds. The structure was changed from a 3D micro-/nano-structure using 10 mm s^{-1} to LIPSS using 1000 mm s^{-1}.

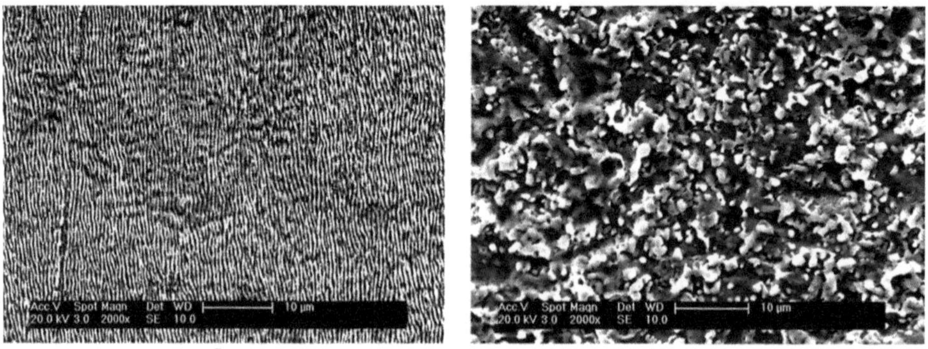

Ps   Ns

Figure 9.7. Changing the stainless steel surface structure as a result of changing the pulse duration LIPSS structure using a picosecond and porous or mesoporous structure using a nanosecond pulse at 50 mm s^{-1} in water.

The effect of pulse duration is shown in figure 9.7. The microstructure was LIPSS using a picosecond laser while it was a porous structure using a nanosecond laser.

9.3.1 Laser melting

Laser melting (LM) is a process in which the laser beam scans over the material's (powder, thin films and bulk material) surface to melt a thin layer of material depending on the scanning and laser parameters. The molten surface directly solidifies due to the high cooling rates, forming different surface structures and topographies. The materials could be metals, polymers or ceramics. This process was first developed at the University of Texas [44]. Laser melting is a complex process due to its fast laser scanning and material transformations in a short period of time. Due to the laser beam shape, the temperature field during laser melting was found to be inhomogeneous resulting in inhomogeneous surface structuring. It has been reported that there is a significant effect of the temperature evolution in laser melting on the quality of the surface including dimensions,

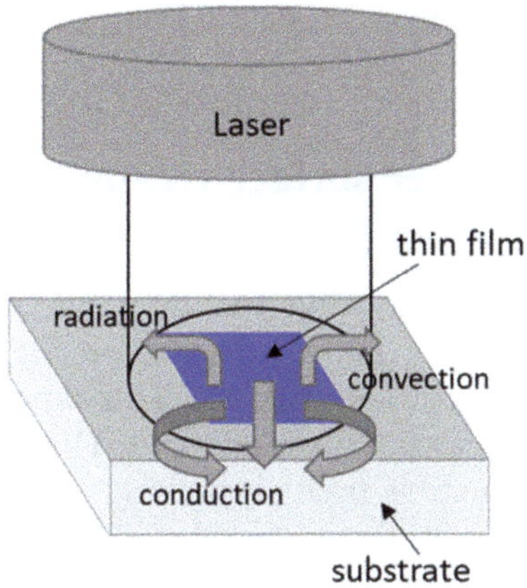

Figure 9.8. Heat transfer during laser sintering of DSSCs.

density, microstructure, mechanical properties, etc. For example, in metals, large thermal gradients increase deformation and residual stresses which may lead to crack formation in the fabricated surface. Moreover, thermal distortion is one serious problem in LM that need to be avoided by understanding the mechanism of the process and the effects of the laser processing parameters [45].

In DSSCs, using laser melting has many advantages over conventional melting methods in terms of the selectivity, scalability, localised machining, cost effectiveness, remote surface interaction, and the applicability to machine different substrates [37]. Figure 9.8 shows the LM principle in DSSCs. During this process, the laser heats the top of the cell affecting the thin layer of the TiO_2 bed. The heat transfer mechanism, in this case, relies on the nanoparticles' bed radiation, convection between the nanoparticles' bed and the environment, and heat conduction inside the nanoparticles' bed and between the nanoparticles' bed and the substrate. The LM process results in the nanoparticles' bed phase change and the corresponding change of the thermal characteristics [45].

For laser melting applications such as laser melting of DSSCs, it has been reported that different lasers can be used such as a CO_2 laser, excimer laser and fibre laser. A CO_2 laser produces a far-infrared radiation with 10.6 μm. It operates either in CW or pulse mode. Both types can be used for the sintering process [46]. An excimer laser produces a UV radiation with a wavelength ranging from 157 to 351 nm. A fibre laser, on the other hand, produces a near-infrared radiation with a wavelength of 1070–1080 nm and it operates in CW and pulse modes. Fibre lasers generate a good beam quality and a very low divergence, which help the users to utilise small spot diameters for precise applications [47].

As mentioned before, when the laser interacts with the substrate, its absorbed energy results in modifying the surface [48]. Depending on the laser characteristics and the material properties, material removal and surface modification occur due to either the photothermal process or photochemical process. It has been reported that the use of CO_2 and fibre lasers lead to photothermal decomposition (or melting) of the material surfaces, while excimer lasers modify the surface due to photochemical decomposition [49–51].

9.4 Previous works on laser surface treatment for application of DSSCs

As mentioned before, after depositing the thin film on the substrate, thermal treatment is required to provide good adhesion on the substrate and to remove the binder and form the mesoporous structure. This can be performed using several methods and laser sintering is one of these methods. Using a laser as a heating source provides some advantages such as time saving and localised processing. There are several works that have focused on studying the effect of a laser as a heating source for solar cell applications. Kim *et al* used the pulsed deposition technique to deposit 30 nm TiO_2 on FTO-coated glass substrate. They used a KrF excimer laser of 248 nm wavelength and 2 J cm^{-2} laser fluence. This process was followed by a laser direct writing technique using Nd:YVO_4 (λ = 355 nm, laser fluence = 0.1 J cm^{-2}, spot size = 250 μm) to form a mesoporous TiO_2 structure. They found that with increasing the mesoporous TiO_2 thin film thickness from 5 to 15 μm, the power conversion efficiency (PCE) was increased from 3.4% to 4.3%. However, with further increasing the thin film layer to 20 μm, the PCE dropped to 3.3%. Thus, 4.3% was the optimum PCE achieved. This was comparable to 4.73% efficiency that was achieved using the conventional method [20]. However, the method mentioned in this work was not simple because they used two kinds of lasers for the depositing and transferring processes, which means that this process is not useful for practical applications. Subsequently, the same group used quasi-continuous wave Nd:YVO_4 (λ = 355 nm, laser fluence = 0.1 J cm^{-2}) in another work to produce a mesoporous TiO_2 film of 12 μm thickness. This work was also performed using one laser for two steps of processing, one for depositing the layer of TiO_2 and one for sintering and producing the mesoporous structure. Low efficiency of 1.84% was achieved compared to 4.73% of the standard oven sintering method (450 °C, 1 h). However, it was higher than the efficiency achieved without the sintering process, which is 0.2% [21]. Here, the authors simplified the processing steps by using one laser for deposition and sintering. However, this method is also not considered preferable for commercialisation. Enhancement of the PCE was achieved by the same group in another work by machining the sample using Nd:YVO_4 (λ = 355 nm, repetition rate = 30 kHz, pulse duration = 60 ns, beam diameter = 25 μm, scanning speed = 100 mm s^{-1}, and laser power = 0.3 W) from the back side of the glass substrate instead of machining from the front side. The laser is mainly absorbed by the TiO_2 layer with little absorption by the FTO-coated glass substrate. This process resulted in melting the TiO_2 nanoparticles forming a continuous thin TiO_2 layer. This thin layer had a thickness of 12 μm and helped in bridging the gaps between electrodes and increasing the PCE to 11.2% [26].

Mincuzzi *et al* used an Nd–YVO_4 laser (with the characteristics of pulse width = 10 ns, wavelength = 355 nm and pulse repletion rate = 30 kHz) for sintering a thin film of TiO_2 that deposited on ITO-coated glass substrate. They depended on manipulating the laser fluence and the focusing distance. They found that the PCE was proportional to the laser fluence. A thickness layer of 8 μm and a PCE of 5.2% was achieved using 500 J cm^{-2}, which was slightly higher than that achieved using the conventional oven sintering process (4.8%), while the non-sintered cell showed a PCE of 0.5% [37]. This work was followed by other work by the same group in which Nd:YVO_4 laser pulses of 355 nm wavelength, 30 kHz repetition rate and 60 ns pulse duration were used to sinter a deposited TiO_2 thin film on ITO-coated glass substrate. They succeeded in fabricating uniform large area (16 cm^2) DSSCs. They based their work on studying the effect of changing the laser power (1, 3, 5 and 7 W) on the PCE of the solar cell. The authors achieved an optimum efficiency of 7.1% after sintering an 8 μm film thickness using 3 W laser power. They found that 3 W was the optimum laser power for heating the surface up to 450 °C, which is required for optimum TiO_2 sintering [23]. This work and the work mentioned in [26] were considered as breakthroughs in DSSCs as the achieved PCE was high.

Another work reported was that conducted by Pan *et al.* Here, an excimer laser was used to produce a 3D mesoporous TiO_2 structure on glass and plastic substrates. The optimum achieved PCE was 3.8% using 80.9 mJ cm^{-2}, which was lower than that achieved using the standard oven sintering process (4.4%) with a film thickness of 5.96 μm [22]. As the processing depth of the excimer laser was low, the process was performed layer by layer for several time processes to achieve a 3D mesoporous TiO_2 structure, which is also not suitable for commercialisation. Yoon *et al* used pulsed Nd:YAG ($\lambda = 1064$ nm, pulse duration = 6 ns, 10 Hz, beam size = 0.9 cm) to sinter a TiO_2 film on FTO-glass substrate. It was noticed with changing the pulse energy, that the PCE was also changed with optimum achieved PCE of 7.62% and 9.89 μm film thickness using 150 mJ pulse energy [25].

Di Carlo *et al* introduced for the first time in a patent, the use of IR and Vis laser light for enhancing the optical properties of TiO_2 paste by adding pigments. After preparing the TiO_2 paste, the samples were irradiated by lasers (Vis or IR), which was followed by UV irradiation for 10 min. It had been found that a high efficiency of 5.5% could be achieved using 532 nm green laser sintering of TiO_2 thin film followed by UV lamp irradiation. This efficiency was similar to that achieved using standard furnace sintering by the same group. However, after 10 min UV irradiation and without laser sintering, the maximum achieved PCE was only 2.2%. The authors suggested that the IR and green laser light sintering was essential for removing the organic binder and achieving high efficiency as a 10 UV lamp illumination was not enough to remove the binder [24].

Ming *et al* reported a novel method of sintering of TiO_2 to produce a mesoporous structure on plastic substrate for DSSCs without damaging the plastic substrate. A Nd:YAG laser was used (with specifications of $\lambda = 1064$ nm, laser fluence = 52 J cm^{-2}, spot size = 368 μm). The machining was performed out of focus using line by line scanning. The film thickness was 13 μm and the optimum PCE achieved was 5.7%, which was comparable to that achieved using the normal sintering method [52]. It was the first time using IR laser sintering to produce high efficiency without the use of UV light radiation as reported before [24].

Recently, Hadi *et al* used CW 1070 nm, Nd:YVO_4 fibre laser with duty cycle 100 ms(ON)/50 ms(OFF) to sinter a TiO_2 thin film layer and formed a mesoporous TiO_2 structure on FTO-coated glass substrate for DSSCs. They studied the effect of laser power density on the film structure and the performance of the DSSCs. The optimum efficiency achieved was around 3% using 85 W cm^{-2}, 60 s irradiation, which was equivalent to that achieved from the standard furnace sintered at 450 °C for 30 min. They also found that by increasing the film thickness to 9 μm, the PCE was increased to 4% [53]. Generally, it can be said that the use of a high efficiency laser of long wavelength increased the economic benefits.

For further enhancement of the solar cell efficiency, some works have been reported using UV and near-infrared lasers to texture furnace-sintered thin films. Yoon *et al* improved the efficiency of DSSCs from 6.33% to 7.62% by irradiating an oven-sintered TiO_2 thin film using a Nd:YAG laser of 1064 nm, 10 kHz, 6 ns and 150 mJ cm^{-2} [25]. Pu *et al*, on the other hand, used a KrF laser to modify the roughness of a furnace-sintered TiO_2 thin film for DSSCs. They used 80 mJ cm^{-2}, 10 Hz with 11 200 pulses to create a texture on a TiO_2 film on FTO-coated glass substrate after furnace sintering at 510 °C for 15 min. Using this method, the transparency of the porous structure was reduced, which increased the light trapping between the textured features. This process enhanced the performance of DSSCs by increasing the PCE to 5.59% with 9 μm coated film thickness [54].

Other attempts were conducted to use a laser for synthesis of a nanocrystalline structure, depositing the thin film on the substrate and processing electrodes as well as sealing the devices for solar cell applications [17, 55]. Melhem *et al* applied the laser pyrolysis method to generate TiO_2 nanocrystals on FTO-coated glass using a 1 kW CO_2 laser. They generated a porous structure with a film thickness of 12 nm. They succeeded in reducing the number of processing steps and achieved a high efficiency of 4.13%, which was equivalent to that achieved using the standard method [55]. Wang *et al* also used the laser pyrolysis method to fabricate a TiO_2/MWCNT nanohybrid material for DSSC application. Using the pyrolysis method, they succeeded in fabricating the hybrid material in a single step process with a good PCE of 3.9% [56].

The pulsed laser deposition method was also used with a 248 nm KrF laser [57] and 355 nm Nd:YAG laser [58] for depositing TiO_2 thin film for solar cell applications. A hierarchical TiO_2 or tree-like structure of the thin film was achieved by using the pulsed laser deposition method. The authors used a 248 nm and 10–15 ns KrF laser to heat up a Ti substrate in O_2 atmosphere to deposit a hierarchical TiO_2 structure on FTO-coated glass substrate. It was noticed that the thickness of the film was increased by increasing the time of laser radiation. This process was followed by annealing at 500 °C in air to achieve the anatase phase from an amorphous one. The optimum PCE achieved was 5% for 7 μm film thickness. [57]. However, a compact layer was achieved using a 355 nm Nd:YAG laser with an enhancement up to 30% of the device efficiency in the presence of the blocking layer. The maximum efficiency achieved in the presence of the blocking layer was 6.15% [58]. Table 9.1 summarises the previous works of laser treated DSSCs.

Table 9.1. Previous work on laser treated DSSCs.

Group	Nc–TiO_2 thin film layer thickness (μm)	PEC %	Laser parameters	Reference
Kim *et al* (2004)	15	4.3	KrF excimer (248 nm wavelength, 2 J cm^{-2} laser fluence) + Nd:YVO_4 (355 nm wavelength, 0.1 J cm^{-2} laser fluence, 250 μm, spot size)	[20]
Kim *et al* (2006)	12	1.84	Nd:YVO_4 (355 nm wavelength, 100 kHz repetition rate, 150 W cm^{-2} laser power density, 250 μm, spot size)	[21]
Mincuzzi *et al* (2009)	8	5.2	Nd:YVO_4 laser (λ = 355 nm, laser fluence = 500 J cm^{-2}, repetition rate = 30 kHz, pulse duration = 10 ns)	[37]
Pan *et al* (2009)	6	3.8	Excimer laser (λ = 248 nm, pulse duration = 20 ns, repetition rate = 10 Hz and laser fluence = 80.9 mJ cm^{-2})	[22]
Kim *et al* (2010)	—	11	Nd:YVO_4 (λ = 355 nm, repetition rate = 30 kHz, pulse duration = 60 ns, beam diameter = 25 μm, scanning speed = 100 mm s^{-1}, and laser power = 0.3 W)	[26]
Sauvage *et al* (2010)	7	5	KrF excimer laser (λ = 248 nm, pulse width = 10–15 ns, laser fluence = 4 J cm^{-2})	[57]
Yoon *et al* (2011)	9.89	7.62	Nd:YAG (λ = 1064 nm, repetition rate = 10 kHz, pulse width = 6 ns and laser fluence = 150 mJ cm^{-2})	[25]
Melhm *et al* (2011)	1.75	4.23	CO_2 laser (CW, λ = 10.6 nm, power = 1 k W)	[55]
Pu and Chen (2012)	9	5.59	KrF excimer (laser fluence = 80 mJ cm^{-2}, repetition rate = 10 Hz and no. of pulses = 11 200 pulses)	[54]
Mincuzzi *et al* (2012)	8	7.1	UV Nd:YVO_4 lasers (355 nm wavelength, 60 ns pulse duration, 30 kHz repetition rate, 3 W laser power)	[23]
Ming *et al* (2014)	13	5.7	Nd:YAG (λ = 1064 nm, laser fluence = 52 J cm^{-2}, spot size = 368 μm)	[52]
Di Carlo *et al* (2014)	—	5.5	ViS and IR laser light	[24]

Wang *et al* (2015)	—	3.9	CO_2 laser (CW, λ = 10.6 nm, power = 600 W)	[56]
Sacco *et al* (2015)	—	6.15	Nd:YAG (λ = 355 nm, repetition rate = 4 Hz, pulse width = 6 ns and laser fluence = 4 J cm^{-2})	[58]
Hadi *et al* (2018)	9	4	CW 1070 nm, Nd:YVO_4 fibre laser (85 W cm^{-2}, 60 s irradiation)	[53]

9.5 Summary

In this chapter, three main generations of solar cells were highlighted. The properties, construction and fabrication methods of DSSCs were discussed. Laser surface treatment is an alternative method that can be used to fabricate and process DSSCs due to its advantages of localised and fast processing compared to conventional methods. This chapter reviewed many efforts studying the effect of laser processing on DSSC fabrications. However, it can be concluded that the use of a laser as a tool for modifying DSSCs is still under development, and a deep understanding of the phenomena involved is much needed.

References

[1] Yu H, Zhang S, Zhao H, Xue B, Liu P and Will G 2009 High-performance TiO_2 photoanode with an efficient electron transport network for dye-sensitized solar cells *J. Phys. Chem.* C **113** 16277–82

[2] Dinçer F and Meral M E 2010 Critical factors that affecting efficiency of solar cells *Smart Grid and Renewable Energy* **1** 47

[3] Kouhnavard M, Ludin N A, Ghaffari B V, Sopian K and Ikeda S 2015 Carbonaceous materials and their advances as a counter electrode in dye-sensitized solar cells: challenges and prospects *ChemSusChem* **8** 1510–33

[4] Kim D S, Gabor A, Yelunder V, Upadhyaya A, Meemongkolkiat V and Rohatgi A 2003 String ribbon silicon solar cells with 17.8% efficiency *3rd World Conf. on Photovoltaic Energy Conversion* ed K Kurokawa, L Kasmerski, B McNelis, M Yamaguchi, C Wronski and W Sinks pp 1293–6

[5] Späth M, Sommeling P, Van Roosmalen J, Smit H, Van der Burg N, Mahieu D, Bakker N and Kroon J 2003 Reproducible manufacturing of dye-sensitized solar cells on a semi-automated baseline *Prog. Photovoltaics Res. Appl.* **11** 207–20

[6] Morgan D L, Shines C J, Jeter S P, Blazka M E, Elwell M R, Wilson R E, Ward S M, Price H C and Moskowitz P D 1997 Comparative pulmonary absorption, distribution, and toxicity of copper gallium diselenide, copper indium diselenide, and cadmium telluride in Sprague–Dawley rats *Toxicol. Appl. Pharmacol.* **147** 399–410

[7] Lee C-P, Lin C-A, Wei T-C, Tsai M-L, Meng Y, Li C-T, Ho K-C, Wu C-I, Lau S-P and He J-H 2015 Economical low-light photovoltaics by using the Pt-free dye-sensitized solar cell with graphene dot/PEDOT: PSS counter electrodes *Nano Energy* **18** 109–17

[8] Mathew S, Yella A, Gao P, Humphry-Baker R, Curchod B F, Ashari-Astani N, Tavernelli I, Rothlisberger U, Nazeeruddin M K and Grätzel M 2014 Dye-sensitized solar cells with 13% efficiency achieved through the molecular engineering of porphyrin sensitizers *Nat. Chem.* **6** 242

[9] Watson T, Mabbett I, Wang H, Peter L and Worsley D 2011 Ultrafast near infrared sintering of TiO_2 layers on metal substrates for dye-sensitized solar cells *Prog. Photovoltaics Res. Appl.* **19** 482–6

[10] Maheswari D and Venkatachalam P 2015 Performance enhancement in dye-sensitized solar cells with composite mixtures of TiO_2 nanoparticles and TiO_2 nanotubes *Acta Metall. Sinica (English Lett.)* **28** 354–61

[11] Roy P, Kim D, Lee K, Spiecker E and Schmuki P 2010 TiO_2 nanotubes and their application in dye-sensitized solar cells *Nanoscale* **2** 45–59

[12] Jasim K E 2011 Dye sensitized solar cells—working principles, challenges and opportunities *Solar Cells—Dye-Sensitized Devices* ed L A Kosyachenko (Rijeka: InTechOpen) 172–204

[13] Hagfeldt A, Boschloo G, Sun L, Kloo L and Pettersson H 2010 Dye-sensitized solar cells *Chem. Rev.* **110** 6595–663

[14] Ting Y P 2008 Structuring and functionalisation of titania *PhD Thesis* Massey University

[15] Apostolopoulou A, Karageorgopoulos D, Rapsomanikis A and Stathatos E 2016 Dye-sensitized solar cells with zinc oxide nanostructured films made with amine oligomers as organic templates and gel electrolytes *J. Clean Energy Technol.* **4** 311–5

[16] Jeng M-J, Wung Y-L, Chang L-B and Chow L 2013 Particle size effects of TiO_2 layers on the solar efficiency of dye-sensitized solar cells *Int. J. Photoenergy* **2013** 563897

[17] Mincuzzi G, Palma A L, Di Carlo A and Brown T M 2016 Laser processing in the manufacture of dye-sensitized and Perovskite solar cell technologies *ChemElectroChem* **3** 9–30

[18] Amiri O and Salavati-Niasari M 2015 High efficiency dye-sensitized solar cells (9.3%) by using a new compact layer: decrease series resistance and increase shunt resistance *Mater. Lett.* **160** 24–7

[19] Uchida S, Tomiha M, Masaki N, Miyazawa A and Takizawa H 2004 Preparation of TiO_2 nanocrystalline electrode for dye-sensitized solar cells by 28 GHz microwave irradiation *Sol. Energy Mater. Sol. Cells* **81** 135–9

[20] Kim H, Kushto G, Arnold C B, Kafafi Z and Piqué A 2004 Laser processing of nanocrystalline TiO_2 films for dye-sensitized solar cells *Appl. Phys. Lett.* **85** 464–6

[21] Kim H, Auyeung R, Ollinger M, Kushto G, Kafafi Z and Piqué A 2006 Laser-sintered mesoporous TiO_2 electrodes for dye-sensitized solar cells *Appl. Phys.* A **83** 73–6

[22] Pan H, Ko S H, Misra N and Grigoropoulos C P 2009 Laser annealed composite titanium dioxide electrodes for dye-sensitized solar cells on glass and plastics *Appl. Phys. Lett.* **94** 071117

[23] Mincuzzi G, Schulz-Ruhtenberg M, Vesce L, Reale A, Di Carlo A, Gillner A and Brown T 2014 Laser processing of TiO_2 films for dye solar cells: a thermal, sintering, throughput and embodied energy investigation *Prog. Photovoltaics Res. Appl.* **22** 308–17

[24] Di Carlo A, Mincuzzi G, Brown T M and Reale A 2014 Sintering process of metal oxide based formulations *Google Patents* EP2697810B1

[25] Yoon J, Jin M and Lee M 2011 Laser-induced control of TiO_2 porosity for enhanced photovoltaic behavior *Adv. Mater.* **23** 3974–8

[26] Kim J, Kim J and Lee M 2010 Laser welding of nanoparticulate TiO_2 and transparent conducting oxide electrodes for highly efficient dye-sensitized solar cell *Nanotechnology* **21** 345203
[27] Wang C, Cheng I-C and Chen J-Z 2015 Ultrafast atmospheric-pressure-plasma-jet sintering of nanoporous TiO_2–SnO_2 composites with features defined by screen-printing *ECS J. Solid State Sci. Technol.* **4** P3020–P5
[28] Zama I, Martelli C and Gorni G 2017 Preparation of TiO_2 paste starting from organic colloidal suspension for semi-transparent DSSC photo-anode application *Mater. Sci. Semicond. Process.* **61** 137–44
[29] Ahmadi S, Asim N, Alghoul M, Hammadi F, Saeedfar K, Ludin N A, Zaidi S H and Sopian K 2014 The role of physical techniques on the preparation of photoanodes for dye sensitized solar cells *Int. J. Photoenergy* **2014** 198734
[30] Ito S, Chen P, Comte P, Nazeeruddin M K, Liska P, Péchy P and Grätzel M 2007 Fabrication of screen-printing pastes from TiO_2 powders for dye-sensitised solar cells *Prog. Photovoltaics Res. Appl.* **15** 603–12
[31] Allen M L, Aronniemi M, Mattila T, Alastalo A, Ojanperä K, Suhonen M and Seppä H 2008 Electrical sintering of nanoparticle structures *Nanotechnology* **19** 175201
[32] Herrmann A, Fiedler J, Ehrmann A, Grethe T, Schwarz-Pfeiffer A and Blachowicz T 2016 Examination of the sintering process dependent micro-and nanostructure of TiO_2 on textile substrates *Photonics for Solar Energy Systems VI* ; R B Wehrspohn and A Gombert *Proc. SPIE Photon.* **9898** 98980s
[33] Miyasaka T, Ikegami M and Kijitori Y 2007 Photovoltaic performance of plastic dye-sensitized electrodes prepared by low-temperature binder-free coating of mesoscopic titania *J. Electrochem. Soc.* **154** A455–A61
[34] Hsu P-Y, Lee H-F, Yang S-M, Chua Y-T, Tung Y-L and Kai J-J 2012 Highly efficient quasi-solid state flexible dye-sensitized solar cells using a compression method and light-confined effect for preparation of TiO_2 photoelectrodes *Procedia Eng.* **36** 439–45
[35] Chang H, Yang Y-J, Li H-C, Hsu C-C, Cheng I-C and Chen J-Z 2013 Preparation of nanoporous TiO_2 films for DSSC application by a rapid atmospheric pressure plasma jet sintering process *J. Power Sources* **234** 16–22
[36] Hooper K, Carnie M, Charbonneau C and Watson T 2014 Near infrared radiation as a rapid heating technique for TiO_2 films on glass mounted dye-sensitized solar cells *Int. J. Photoenergy* **2014**
[37] Mincuzzi G, Vesce L, Reale A, Di Carlo A and Brown T M 2009 Efficient sintering of nanocrystalline titanium dioxide films for dye solar cells via raster scanning laser *Appl. Phys. Lett.* **95** 240
[38] Bäuerle D 2000 *Laser Processing and Chemistry* vol 3 (Berlin: Springer)
[39] Von der Linde D, Sokolowski-Tinten K and Bialkowski J 1997 Laser–solid interaction in the femtosecond time regime *Appl. Surf. Sci.* **109** 1–10
[40] Lippert T 2010 *Laser–Surface Interactions for New Materials Production* (Berlin: Springer) pp 141–75
[41] Batteh J, Chen M and Mazumder J 2000 A stagnation flow analysis of the heat transfer and fluid flow phenomena in laser drilling *J. Heat Transfer* **122** 801–7
[42] Brown M S and Arnold C B 2010 *Laser Precision Microfabrication* (Berlin: Springer) pp 91–120

[43] Rajab F H, Whitehead D, Liu Z and Li L 2017 Characteristics of hierarchical micro/nano surface structure formation generated by picosecond laser processing in water and air *Appl. Phys.* B **123** 282

[44] Levy G N, Schindel R and Kruth J-P 2003 Rapid manufacturing and rapid tooling with layer manufacturing (LM) technologies, state of the art and future perspectives *CIRP Ann.* **52** 589–609

[45] Zeng K, Pal D and Stucker B 2012 A review of thermal analysis methods in laser sintering and selective laser melting *Proc. Solid Freeform Fabrication Symposium (Austin, TX)* ed J Beaman, D Bourell, R Crawford, H Marcus and C C Seepersad pp 796–814

[46] Vainos N A 2012 *Laser Growth and Processing of Photonic Devices* (Amsterdam: Elsevier)

[47] Zervas M N and Codemard C A 2014 High power fiber lasers: a review *IEEE J. Sel. Top. Quantum Electron.* **20** 219–41

[48] Modest M, Ready J and Farson D 2001 *Handbook of Laser Materials Processing* (Orlando, FL: Magnolia)

[49] Ready J F 1997 *Industrial Applications of Lasers* (Amsterdam: Elsevier)

[50] Madou M J 2002 *Fundamentals of Microfabrication: The Science of Miniaturization* (Boca Raton, FL: CRC press)

[51] Joya Y F 2011 Titanium dioxide films Prepared by sol–gel/laser-induced technique for inactivation of bacteria *PhD Thesis* (University of Manchester)

[52] Ming L, Yang H, Zhang W, Zeng X, Xiong D, Xu Z, Wang H, Chen W, Xu X and Wang M 2014 Selective laser sintering of TiO_2 nanoparticle film on plastic conductive substrate for highly efficient flexible dye-sensitized solar cell application *J. Mater. Chem.* A **2** 4566–73

[53] Hadi A, Alhabradi M, Chen Q, Liu H, Guo W, Curioni M, Cernik R and Liu Z 2018 Rapid fabrication of mesoporous TiO_2 thin films by pulsed fibre laser for dye sensitized solar cells *Appl. Surf. Sci.* **428** 1089–97

[54] Pu M-Y and Chen J Z 2012 Improved performance of dye-sensitized solar cells with laser-textured nanoporous TiO_2 photoanodes *Mater. Lett.* **66** 162–4

[55] Melhem H, Simon P, Beouch L, Goubard F, Boucharef M, Di Bin C, Leconte Y, Ratier B, Herlin-Boime N and Bouclé J 2011 TiO_2 nanocrystals synthesized by laser pyrolysis for the up-scaling of efficient solid-state dye-sensitized solar cells *Adv. Energy Mater.* **1** 908–16

[56] Wang J, Lin Y, Pinault M, Filoramo A, Fabert M, Ratier B, Boucle J and Herlin-Boime N 2015 Single-step preparation of TiO_2/MWCNT nanohybrid materials by laser pyrolysis and application to efficient photovoltaic energy conversion *ACS Appl. Mater. Interfaces* **7** 51–6

[57] Sauvage F, Di Fonzo F, Li Bassi A, Casari C S, Russo V, Divitini G, Ducati C, Bottani C E, Comte P and Graetzel M 2010 Hierarchical TiO_2 photoanode for dye-sensitized solar cells *Nano Lett.* **10** 2562–7

[58] Sacco A, Di Bella M S, Gerosa M, Chiodoni A, Bianco S, Mosca M, Macaluso R, Calì C and Pirri C F 2015 Enhancement of photoconversion efficiency in dye-sensitized solar cells exploiting pulsed laser deposited niobium pentoxide blocking layers *Thin Solid Films* **574** 38–42

Chapter 10

Laser micro-processing for polymers and silicon for microfluidic applications

Shashi Prakash

Microfluidic devices play an important role in day to day life. Rapid fabrication of these microfluidic technologies/devices has become the need of the hour. Developments in micro-devices heavily rely on microfabrication methods. Lasers play a predominant role as a microfabrication tool. Although lasers have been used for a wide variety of materials, they are particularly and successfully used for processing polymers and silicon within the microfluidic industry. From low cost CO_2 lasers to costly femtosecond lasers, many studies have been conducted on laser processing using these substrate materials. Polymers as well as silicon are used for a number of microfluidic devices in chemical, biological and sensing devices. This chapter focuses on the fundamental principles and various types of laser processes utilized for the micro-processing of polymers and silicon.

10.1 Introduction

Microfluidics and smart micro-devices are evolving day by day, replacing conventional methods of biological and chemical testing involving large amounts of costly reagents and large devices. These less than palm-size devices are able to perform tests/functions in fractions of the time and more reliably than their old counterparts. However, their development lies in easy and cost effective methods of producing them. More often these devices are made portable and disposable. Further, evolution and development of these micro-devices also depend upon innovative changes in design and easy fabrication as simulation often may not provide exact results given the magnitude of miniaturization involved. The most popular microfluidic devices include gas chromatographs, electrophoresis devices, electro-osmotic systems, separation devices, micro-mixers, micro-pumps, bio-analytical devices, micro-reactors, chemical devices, micro heat exchangers, DNA amplifiers, cytometers, etc. Micro-electro-mechanical systems (MEMS) devices also include miniaturized embedded systems,

doi:10.1088/978-0-7503-1683-5ch10

having some micromachined parts enabling higher level functions and are mostly used in control and sensor based applications. Figure 10.1 shows the typical application areas of MEMS related technologies and their advantages.

The earliest description of MEMS devices was found in the late 1960s in the form of microelectronic circuits. With successful inventions in microelectronics, miniaturization was also explored in the area of biological and chemical testing since reagents to be used for such devices were either expensive or rarely available. In the late 1970s, electrophoretic devices were invented as a significant biological invention and also gas chromatographs for sensing of gases [1]. However, most of the recent developments came in the late 1990s with the evolution of inkjet printers. Further, a series of new inventions took place including typical household devices such as pregnancy test kits, diabetic test kits, etc. Many of these devices involve microstructures having sizes in the range of 50–500 μm [2]. In recent developments, many biological test devices can replicate the functions of human organs and different types of testing can be performed [3]. Complex chemical micro-devices are now been able to perform a number of functions and complex tasks which were difficult in larger normal scale devices [4].

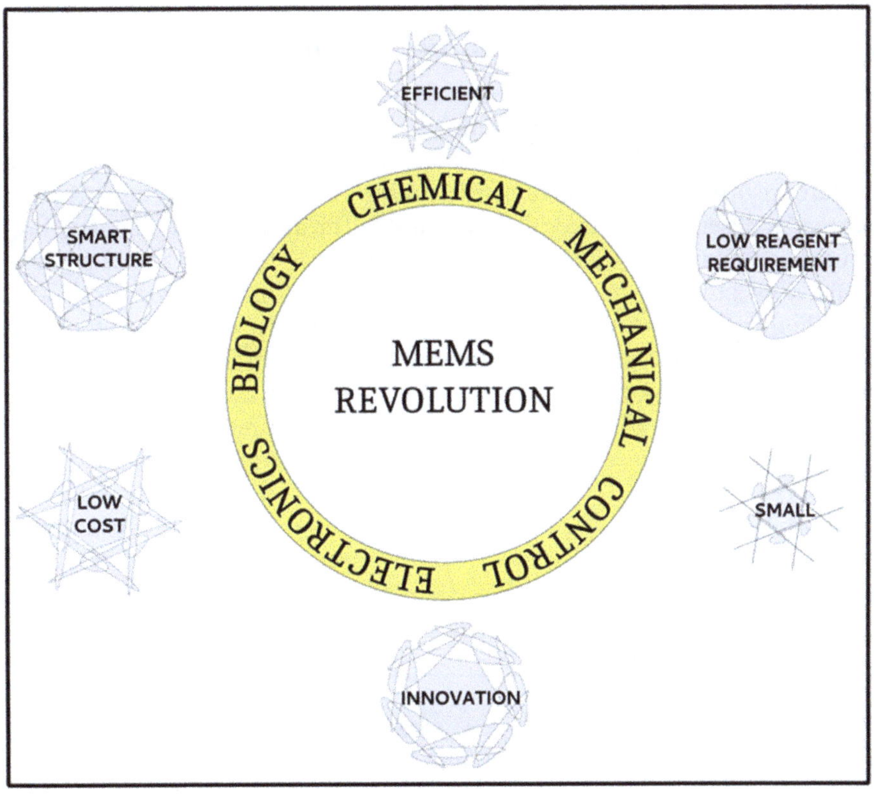

Figure 10.1. MEMS revolution.

In the area of cooling electronic circuitry, Tuckerman and Pease [5] introduced the idea of forced cooling by means of microchannels which formed the basis of the miniaturization of electronic devices.

Although a number of micromachining processes are available for the formation of microstructures, laser based processes offer a unique advantage of low time consumption, ease of operation and almost no material restrictions. Lithography, injection molding, chemical etching, electro-discharge machining, electro-chemical machining, mechanical micro-milling, etc, have been used by different researchers for microstructure formation.

However, laser based processes stand out as one of the most used fabrication processes for micro-devices [6]. Most of such machining systems require clean-room facilities, highly skilled labor, job intensive fixture designs and expensive equipment. On the other hand, lasers generally do not require clean-room facilities, can be operated by semi-skilled personnel, can be operated without any fixture and clean-room fixture and can often be cheaper than other equipment [7]. A number of operations can be performed on a single laser system including drilling, cutting, microstructuring, joining and engraving making it a unified solution for making a microfluidic device in entirety [8].

10.2 Types of laser systems and related ablation phenomenon

Commonly used laser systems for microfabrication can be subdivided into different wavelength categories. As per their change in wavelength, the type and quality of ablation also changes. However, it must be noted that apart from the wavelength, pulse width also plays an important role. Laser systems cannot be generalized and any particular laser system may not be suitable for all materials. Depending upon the wavelength and pulse widths, the cost of a laser system also varies significantly. Normally, low power infrared (IR) lasers cost less than ultraviolet (UV) lasers. Similarly higher pulse width lasers also cost significantly less than low pulse width lasers for the same power/capacity range. The price of laser systems may vary significantly, hence selecting a particular type of laser is a complex issue and must be thought upon carefully. Figure 10.2 shows the most commonly used laser systems at different wavelengths.

CO_2 lasers fall in the category of IR lasers having available wavelengths of 9.3 μm and 10.6 μm. CO_2 lasers having 10.6 μm wavelengths are commercially available in a cheaper price range. Typical 60 W CO_2 laser systems made in China are available in

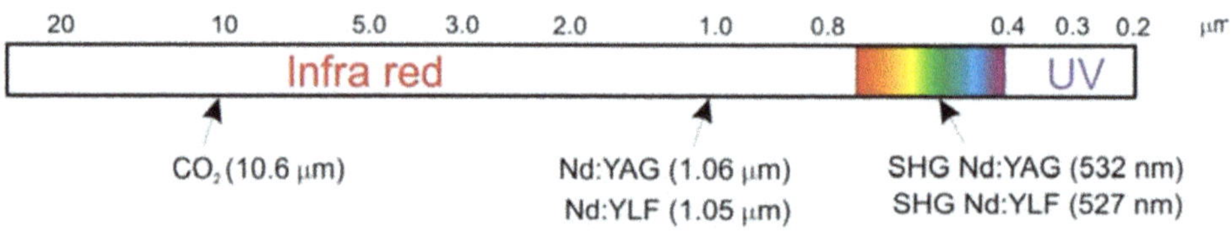

Figure 10.2. Infrared, visible and UV spectral wavelength zones, with the positions of the CO_2, Nd:YAG and Nd:YLF laser wavelengths. SHG denotes the second harmonic generated (double-frequency, half wavelength) versions of the latter two.

the market with prices as low as 1500 USD. However, good quality medium power (< 100 W) CO_2 laser systems can be found in the range of 15 000–20 000 USD. CO_2 laser systems fall in the category of IR lasers where the ablation phenomenon is dominated by thermal ablation. A thermal ablation phenomenon consists of typical heating, melting and vaporization cycle. Due to the dominance of thermal nature, heat related defects predominantly occur in such kind of ablation. The heat affected zone, charring, melting, clogging, bulging and presence of burr are some of the common accompanying defects with such lasers. However, they can be well controlled for certain materials requiring less energy to vaporize. Due to this reason, CO_2 lasers are mostly preferred for polymers and organic materials. It must also be noted that 10.6 μm of wavelength may not suit many materials as reflectivity may play an important role. Hence, the absorptivity of the material at this particular wavelength also plays an important role. Figure 10.3 shows the reflectivity and transmissivity of polymethyl methacrylate (PMMA) at different wavelengths. At 10.6 μm wavelength (i.e. CO_2 laser's wavelength), PMMA was found to have reflectivity of 4.67% and transmissivity of 0.31%. This means that PMMA absorbs nearly 95% of the beam radiation emitted by a CO_2 laser beam. Considering such a small amount of power loss (5%), the CO_2 laser beam can be considered as one of the foremost solutions for microstructuring on such a material. However, absorptivity may not be a sufficient criterion for selecting a particular type of laser for different materials. A material's melting behavior, energy requirement during heating, melting and vaporization also play key roles. In general, a lower amount of latent heat of melting results in a lower amount of charring or heat related defects. A lower amount of energy requirement also causes a lower heat affected zone and results in a clean cut at the edges along with smoother surface walls.

A typical material removal cycle of PMMA when subjected to CO_2 laser processing is given in figure 10.4. Here, T_g denotes the glass transition temperature, T_m denotes the melting temperature, T_{ds} and T_{de} signify decomposition start and decomposition end temperature. In the case of a simple microchanneling operation, depth and cut section profile in a single pass can be easily determined using energy

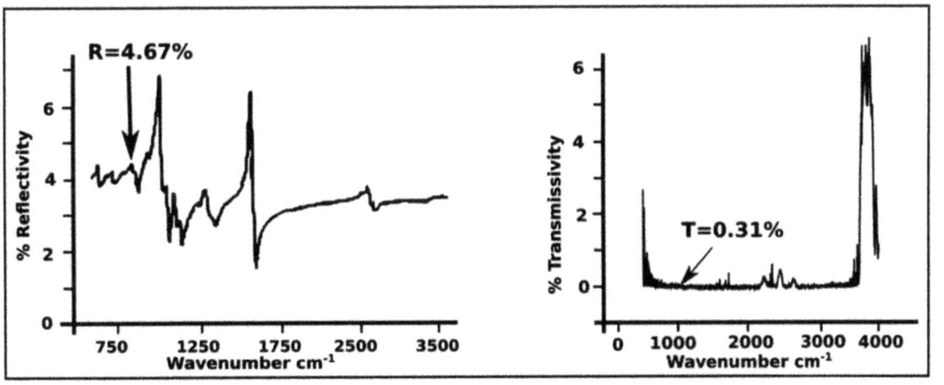

Figure 10.3. Reflectivity and transmissivity of PMMA at different wavelengths.

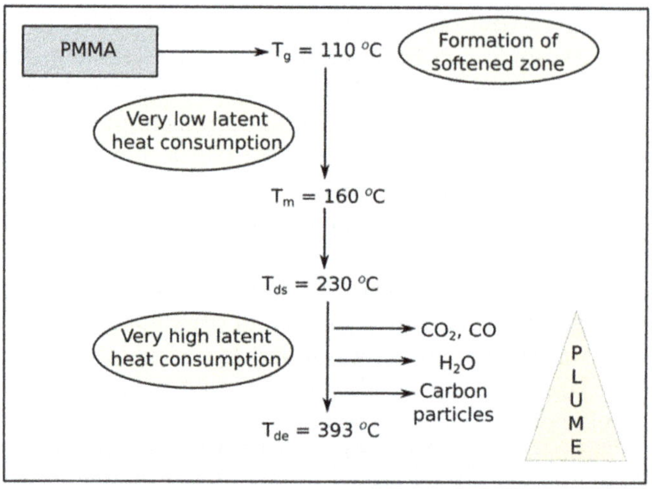

Figure 10.4. Material removal process of PMMA due to a thermal cycle.

balance as given by [9] for PMMA. The cut section profile can be obtained using equations (10.1) and (10.2).

$$Z_{\max} = \frac{\alpha\sqrt{2}}{w\rho\sqrt{\pi}(c_p\Delta T + H_L)} \times \frac{P}{U} \tag{10.1}$$

$$Z = Z_{\max} e^{-2(y^2/w^2)} \tag{10.2}$$

where z represents depth corresponding to each y point, α = laser beam absorptivity, ρ = material density, P = average laser power, U = scanning speed, $c_p\Delta T + H_L$ = enthalpy of vaporization and w = beam spot diameter.

Most of the IR laser ablation processes follow similar material removal phenomenon. In metals, the conduction loss is more compared to polymers, hence equation (10.1) may not predict correct results as the equation does not consider the conduction energy losses.

Pulsed IR lasers such as Nd:YAG lasers having a wavelength of 1.064 μm also ablate the material following a thermal cycle, i.e. heating, melting and vaporization. However, due to pulsing, the beam contact time with the substrate material is considerably low resulting in lower heat related defects compared to continuous or quasi-continuous CO_2 lasers [10]. IR lasers having 1.064 m may not be as suitable for many polymeric substrates as CO_2 lasers due to low absorptivity. That is why such IR lasers are mostly used with metals as striking energy density and absorptivity are too high. As we lower the wavelengths, second harmonic lasers or frequency doubled Nd:YAG lasers (532 nm) are also used in the material processing industry. However, their use is comparatively low due to complex systems and maintenance requirement. For polymers, frequency doubled Nd:YAG lasers result in better quality cuts. The output heat affected zone is also lower than 1.06 m Nd:YAG lasers and 10.6 μm CO_2 lasers.

Third harmonic Nd:YAG lasers, also called frequency tripled Nd:YAG lasers, and excimer lasers fall into the category of UV lasers having wavelengths in the range of 351–147 nm. UV lasers have extremely intense pulses which are sufficient to break polymeric bonds [11]. In other words, these lasers ablate the material predominantly by breaking the bonds between atoms leaving an even smaller heat affected zone and fewer related defects. This particular type of ablation is also referred to as cold ablation. It must be noted though that they have a high energy pulse and therefore maintenance is a severe problem. This is also the reason why these lasers are limited to polymer substrates only as metals require more energy and more energy leads to more maintenance requirement. Also, the processing time is significantly higher than higher wavelength lasers.

Ultrashort lasers are recent developments in the area of laser micro-processing where beam interaction time lies in the order of picoseconds to femtoseconds. Here, material ablation is also predominantly cold, similar to UV lasers, and heat related defects are almost invisible. The material removal phenomenon generally takes place via two methods, i.e. multiphoton absorption or Coulomb explosion depending upon the beam intensity. Both these methods involve the breaking of bonds by either simultaneous photon absorption or atom stripping (also called Zener tunneling) [12]. Ablation quality in such a type of system is extremely good and the tolerance up to the nanometer range can be easily attainable [13]. However, these systems are much costlier and, therefore, they are not used as much as other systems despite leading to high quality end results.

10.3 Laser micro-processing for silicon and polymers

Over the years a number of materials have been used for micro-processing. Polymers are mainly used in biological and chemical applications while metal or composite substrates are mostly used in mechanical, cooling or nuclear applications. Often, lasers have been primarily used for polymers and silicon as microfluidic substrate material. This section mainly emphasizes the previous works and important observations during different types of laser machining on these substrate materials.

Many authors have used laser processed polymers for microfluidic applications. Polydimethyl siloxane (PDMS) is very commonly used for many microfluidic devices as it possesses high thermal stability, mechanical strength and excellent optical transparency. It has been used for micro-reactors, electrophoresis devices, separation devices, etc [14]. Although polydimethyl siloxane based mixing devices have been fabricated earlier for lower flow rate devices, the manufacturing complexity remains a significant problem [15]. CO_2 lasers have not been directly used for microstructure fabrication on PDMS as this results in a bulk of charring at the outer edges making it an inefficient process. But, CO_2 lasers are used for microstructure fabrication on PDMS in indirect ways. For example Arun *et al* [15] fabricated a paper PDMS microchannel in which the channel height was kept as small as 0.16 mm. The channel was fabricated using a PDMS mold created using soft lithography methods. The negative template was fabricated by a CO_2 laser. It was observed that up to a limited depth (typically in the range of 100 μm), CO_2 lasers can be used successfully for PDMS. An alternative method of producing microchannels on PDMS by a CO_2 laser was given by Liu and Gong [16], as shown in figure 10.5. The through cut pattern transfer (TCT) method was used where a glass or PDMS base was used bonded to cut a PDMS section for a clear and frictionless base of the microchannels.

Ultrafast lasers have been proved to be most effective lasers for PDMS processing. Due to lower beam diameter and cold form of ablation the cuts are relatively cleaner than CO_2 lasers without resulting in charring at the edges [17]. Polycarbonate is another important transparent polymer used for microfluidic devices. Polycarbonate is mechanically stronger than PMMA, more scratch resistant and possesses a higher glass transition temperature of 145 °C [18]. Fabrication of microstructures on polycarbonate using a CO_2 laser was also found to deliver a

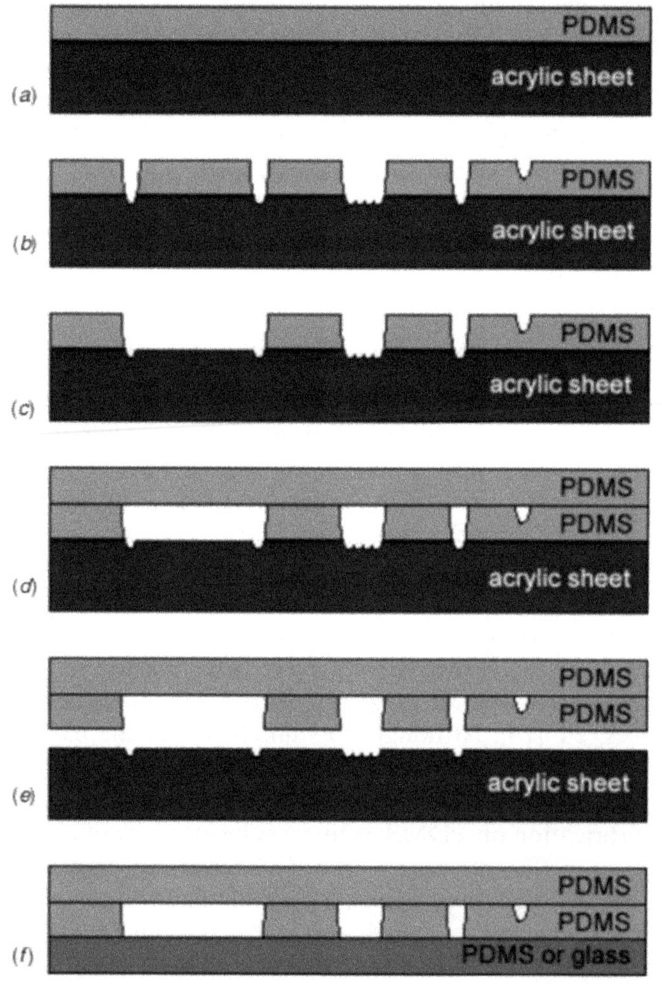

Figure 10.5. Microchannel fabrication by a through cutting and pattern transfer method as suggested by Liu and Gong [16].

larger heat affected zone. A thick molten layer was formed on the microchannel walls as shown in figure 10.6. Ultrafast lasers have been successfully used for fabricating microfluidic structures on polycarbonate with smooth walls and sub-micron accuracy [19].

PMMA is probably the most used polymer for microfluidic devices due to its ability to get a clean cut using cheaper CO_2 lasers. However, it must be noted that CO_2 laser processing of PMMA is absolutely defect free. Burrs, clogging, heat affected zone and proper dimensional control are few of the commonly occurring defects in this processing. In order to make the smooth microchannel walls, multiple-pass processing [21] and defocused processing [22] are commonly known and utilized. Multi-pass processing helps to lower the amount of energy deposition on material in each pass thus lowering the heat related defects. Figure 10.7 shows the microchannel structure after each pass up to seven numbers of passes. Smoothing at the microchannel surface can be clearly seen in these SEM images. Similarly, defocused processing also allows a lower amount of energy intensity to interact with material, thus resulting in improved surface roughness. Figure 10.8 shows the three-dimensional optical images of microchannel processes at defocused distances of 1, 3 and 5 mm [22].

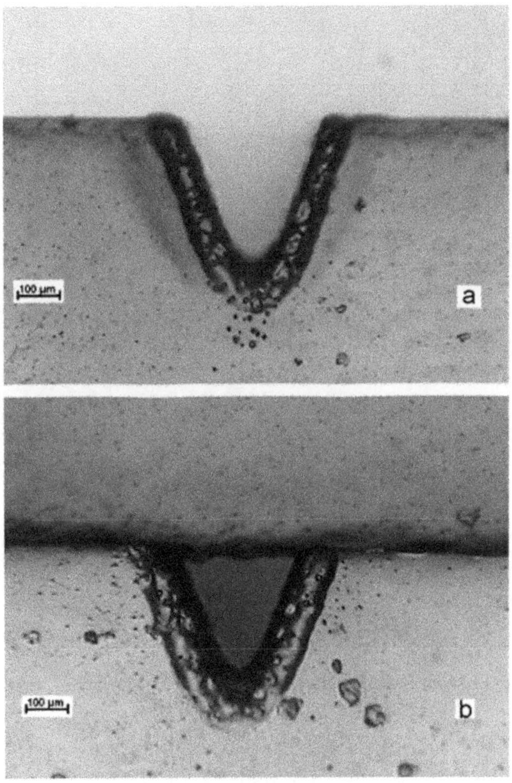

Figure 10.6. Micromachining of polycarbonate with a CO_2 laser (a) before bonding and (b) after bonding. Reprinted from [20], copyright (2009), with permission from Elsevier.

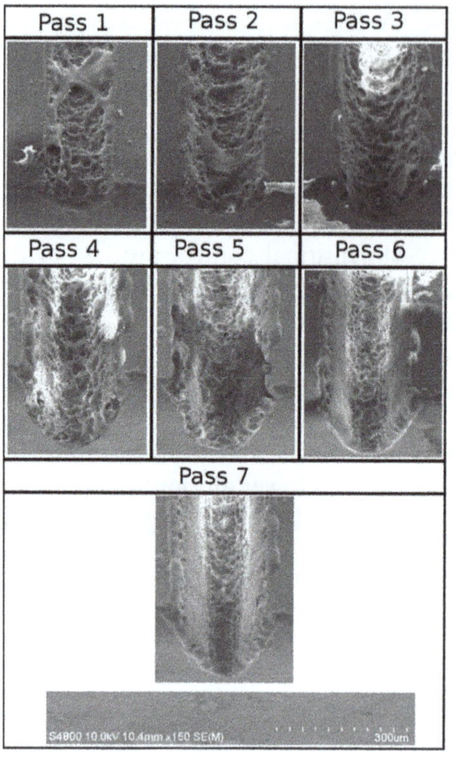

Figure 10.7. Microchannel structure in different numbers of passes. Reprinted from [21], copyright (2017), with permission from Elsevier.

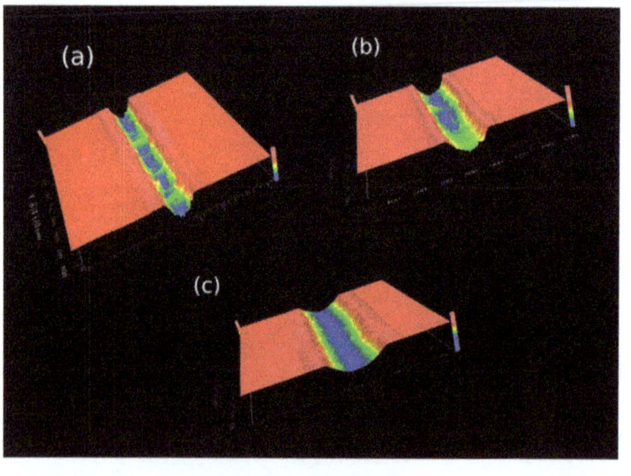

Figure 10.8. Microchannels processed at different amounts of defocusing [22].

Recently, mask based techniques were also developed for fabricating rectangular cross-section microchannels on PMMA using CO_2 lasers [23]. A schematic diagram for producing rectangular cross-section microchannels is given in figure 10.9 utilizing masks prepared using fiber lasers. PMMA is also found to be precisely processed

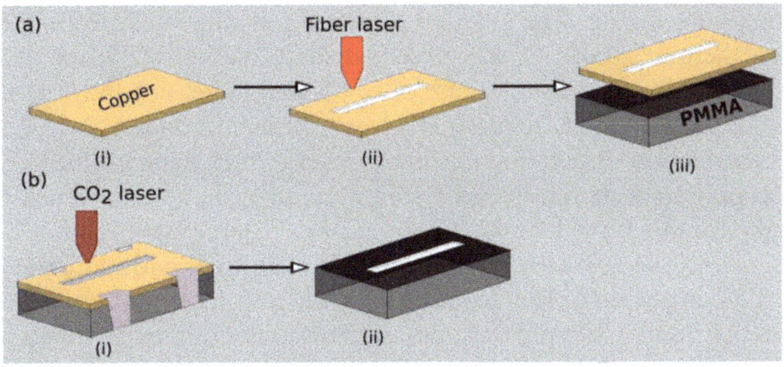

Figure 10.9. Schematic diagram for fabricating rectangular microchannels on PMMA using CO_2 lasers. Reprinted from [23], copyright (2017), with permission from Elsevier.

under ultrashort lasers [24] without producing significant thermal damage on the microchannel rims.

Other than polymers, silicon based microstructures have been largely processed by lasers. Silicon is primarily used as a base material for printed circuit boards, solar cells [25], electronic circuits, sensing devices [26], optical wave guides [27] and heat transfer based applications. Conde *et al* [28] discussed the microstructure evolution on laser ablated silicon and presented theoretical models. The authors demonstrated that initial morphology has a key role in determining the ablation effects due to the small variations of the geometries of the target substrates.

Although it is well known that CO_2 laser beams do not absorb well with silicon, Chung *et al* [29] presented a novel concept of using glass at the back of the silicon for CO_2 laser processing. Though the mechanism for this change in absorption pattern was not clear, significantly improved results were obtained for silicon ablation. The process required multiple numbers of passes ranging from 10 to 80, while material removal rate was observed to be comparatively lower.

In most of the available literature, nanosecond Nd:YAG or fiber lasers have been commonly used for silicon. Lim *et al* [30] fabricated herringbone structures utilizing Nd:YAG lasers on 660 μm thick silicon substrate with widths of 250, 200, 160 and 125 μm and depth of 125 μm. Defocused processing or chemical etching after laser processing was employed for achieving higher surface quality structures when machined using nanosecond lasers. O'Neil and Li [31] demonstrated that with the application of master oscillator power amplifier-based high-brightness Yb-based fiber lasers, microstructures can be formed on silicon with minimum thermally induced defects such as thermal crack, recast layers and surface damage. These lasers can be operated at 30–50 ns of pulse width and 500 kHz of frequency leading to high surface finish at the microstructure edges. While the peak intensity can be increased to as high as 2 GW cm^{-2}, the etching rate was observed to be 230 000 000 $\mu m^3\ s^{-1}$ for a 20 W fiber laser.

UV lasers have also been used for silicon micro-processing. Although a lower etching rate has always been a problem, the quality of cut remains superior to nanosecond lasers. Chen and Darling [32] used a frequency tripled Nd:YAG laser (UV range) for silicon micromachining. The ablation phenomenon was found to be predominated by the photo thermal nature. It was observed that ablation efficiency depends upon threshold efficiency for silicon in the case of UV lasers. Increasing the laser fluence does not result in improvement of ablation efficiency. In fact, reducing the scanning speed may result in reduction of ablation efficiency, as shown in figure 10.10 [32].

Similar to polymers, femtosecond lasers were also found to produce sub-micron level accuracy for silicon substrates. Kam *et al* [33] demonstrated that by applying 20 s of etching as a post-processing process, microchannels with different depths can be produced having sub-micron accuracy as shown in figure 10.11.

Ngoi *et al* [35] used a 400 nm Ti:sapphire femtosecond laser having a pulse width of 150 fs to etch the nano-channels on silicon having sizes as small as 100 nm depth and 500 nm width. Tran *et al* [36] also studied the sub-surface damage by a femtosecond Ti:sapphire laser on silicon substrates using sub-threshold pulses. It took five number of pulses before any damaged surface appeared on the silicon substrates. Amer *et al* [37] compared the femtosecond laser and nanosecond laser machining on single crystal silicon. For this purpose, the authors used a 150 fs laser compared to a 30 ns laser. It was observed that both types of lasers resulted in

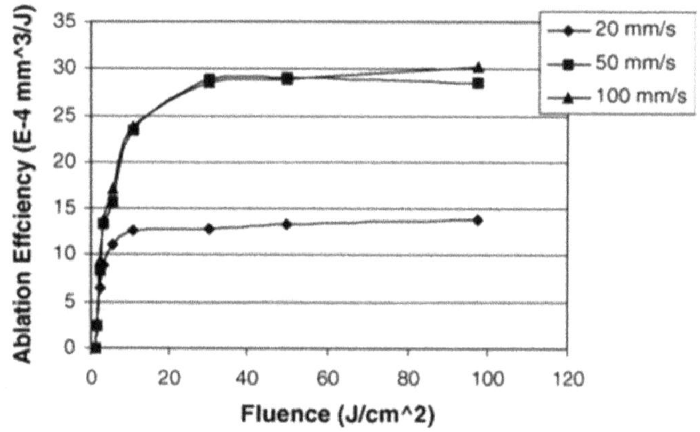

Figure 10.10. The laser ablation efficiency vs laser fluence for silicon with different cutting speeds. Reprinted from [32], copyright (2005), with permission from Elsevier.

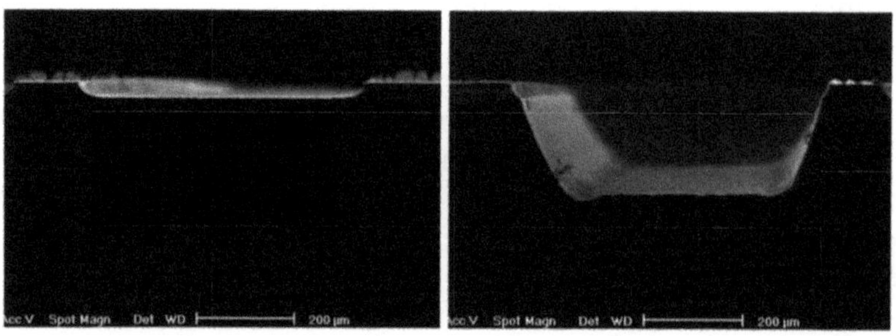

Figure 10.11. SEM images of the cross-section of femtosecond laser machined microchannels onto silicon after 20 s of acid etching [34].

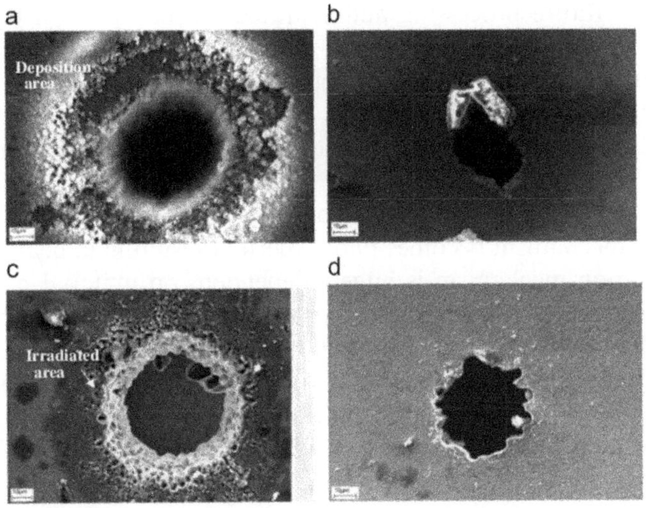

Figure 10.12. Entrance and exit holes for air and water drilling at a scan speed of 0.5 mm s^{-1} (a) entrance hole, (b) exit hole in air drilling, (c) entrance hole, and (d) exit hole underwater drilling. Reprinted from [38], copyright (2011), with permission from Elsevier.

induced stress in the substrate material, which was measured using Raman spectroscopy. However, the stress induced by the nanosecond laser was larger than the femtosecond lasers. Both types of lasers were found to create amorphization up to 25% in single crystal silicon.

Apart from direct laser writing on silicon substrates, a number of studies are available where the processing was performed under different environments. The purpose of using different surrounding environments has been to reduce heat related defects, redeposition and burr formation. Underwater nanosecond laser processing was found to bring considerable improvement in the form of reduction of burrs and debris from microchannel edges. Wee *et al* [38] designed an underwater processing set-up for laser microdrilling on silicon. Water was allowed to flow at a fixed flow rate of 2 l min^{-1}. Flowing water reduces the material redeposition at the upper surface and a clean surface around the hole was obtained, as shown in figure 10.12.

Tangwarodomnukun *et al* [39] utilized a combination of laser heating and water jet to remove material from silicon. It was observed that the high-pressure water jet can substantially reduce thermal damage from the cutting edges. Since the material was heated well below its melting temperature, this technique relies on shear stress formation in the heating zone. In this technique, an off-axial jet having pressure up to 67 MPa with nozzle diameter of 570 μm was utilized along with a nanosecond pulsed fiber laser. In a slightly different technique by Li *et al* [40], a water jet guided laser beam was used, with the laser beam and water jet having coaxial orientation. Both techniques resulted in minimized thermal defects at the microchannel edges.

10.4 Future challenges

Laser processing presents great possibilities for micro-/nano-processing of different material substrates. A number of materials have already been explored and a lot more remain for future processing and characterization. Though this chapter has been dedicated to polymers and silicon due to the wide availability of related publications, a number of studies on other materials are also available in the literature. As it has been widely said, the laser is a solution to many yet unknown problems. Still, there are some significant challenges that need to be addressed as noted below:

1. While the CO_2 laser and nanosecond lasers are available at lower cost, the ablation mechanism becomes photo thermal resulting in heat related defects.
2. The ablation mechanism is hugely dependent on material properties. Since different materials have different properties, it becomes invariably different for each material. Simulation often leads to incorrect results and prediction of the actual response becomes a tough job. Experimentation remains the most viable option for process development. Also, analytical models for one material may not be suitable for other materials due to different material properties. Hence, different analytical models may be required for different materials.
3. Although ultrashort lasers such as femtosecond lasers and picosecond lasers offer significant improvement in ablation quality, the cost, maintenance and operation time is still a big concern.
4. Surface roughness of the laser ablated zone may not be uniform and well below the acceptable limit for many materials and for many applications resulting in difficulty in applying secondary post-processing operations.
5. Although most lasers can produce defined microstructures, producing a high aspect ratio is tough and may require multiple processing steps.

References

[1] Terry S, Jerman J and Angell J 1979 A gas chromatographic air analyzer fabricated on a silicon wafer *IEEE Trans. Electron Dev.* **26** 1880–6

[2] Yuan D and Das S 2007 Experimental and theoretical analysis of direct write laser micromachining of polymethyl methacrylate by CO_2 laser ablation *J. Appl. Phys.* **101**

[3] Bhatia S N and Ingber D E 2014 Microfluidic organs-on-chips *Nat. Biotech.* **32** 760–72

[4] Tanaka Y, Sato K, Shimizu T, Yamato M, Okano T and Kitamori T 2007 Biological cells on micro-chips: new technologies and applications *Biosens. Bioelectron.* **23** 449–58

[5] Tuckerman D and Pease R 1981 High-performance heat sinking for VLSI *IEEE Trans. Electron Dev.* **2** 126–9

[6] Becker H and Locascio L E 2002 Polymer microfluidic devices *Talanta* **56** 267–87

[7] Fu L-M, Ju W-J, Yang R-J and Wang Y-N 2012 Rapid prototyping of glass-based microfluidic chips utilizing two-pass defocused CO_2 laser beam method *Microfluid. Nanofluid.* **14** 479–87

[8] Malek C G K 2006 Laser processing for bio-microfluidics applications (part II) *Anal. Bioanal. Chem.* **385** 1362–9

[9] Prakash S and Kumar S 2015 Profile and depth prediction in single-pass and two-pass CO_2 laser microchanneling processes *J. Micromech. Microeng.* **25** 035010

[10] Banerjee A, Ogale A A, Das C, Mitra K and Subramanian C 2005 Temperature distribution in different materials due to short pulse laser irradiation *Heat Transfer Eng.* **26** 41–9

[11] Molian P, Pecholt B and Gupta S 2009 Picosecond pulsed laser ablation and micromachining of 4H-SiC wafers *Appl. Surf. Sci.* **255** 4515–20

[12] Joglekar A P, Liu H-H, Meyhöfer E, Mourou G and Hunt A J 2004 Optics at critical intensity: applications to nanomorphing *Proc. Natl. Acad. Sci. U.S.A.* **101** 5856–61

[13] Sokolowski-Tinten K, Bialkowski J, Cavalleri A, von der Linde D, Oparin A, Meyer-ter Vehn J and Anisimov S I 1998 Transient states of matter during short pulse laser ablation *Phys. Rev. Lett.* **81** 224–7

[14] Fujii T 2002 PDMS-based microfluidic devices for biomedical applications *Microelectron. Eng.* **61–62** 907–14

[15] Arun R K, Priyadarshini N, Chaudhury K, Chanda N, Biswas G and Chakraborty S 2016 Paper-PDMS hybrid microchannel: a platform for rapid fluid-transport and mixing *J. Micromech. Microeng.* **26** 105008

[16] Liu H-B and Gong H-Q 2009 Templateless prototyping of polydimethylsiloxane microfluidic structures using a pulsed CO_2 laser *J. Micromech. Microeng.* **19** 037002

[17] Darvishi S, Cubaud T and Longtin J P 2012 Ultrafast laser machining of tapered microchannels in glass and PDMS *Opt. Lasers Eng.* **50** 210–4

[18] Ogonczyk D, Wegrzyn J, Jankowski P, Dabrowski B and Garstecki P 2010 Bonding of microfluidic devices fabricated in polycarbonate *Lab Chip* **10** 1324–7

[19] Zheng H, Liu H, Wan S, Lim G, Nikumb S and Chen Q 2006 Ultrashort pulse laser micromachined microchannels and their application in an optical switch *Int. J. Adv. Manuf. Technol.* **27** 925–9

[20] Qi H, Chen T, Yao L and Zuo T 2009 Micromachining of microchannel on the polycarbonate substrate with CO_2 laser direct-writing ablation *Opt. Lasers Eng.* **47** 594–8

[21] Prakash S and Kumar S 2017 Experimental investigations and analytical modeling of multi-pass CO_2 laser processing on PMMA *Precis. Eng.* **49** 220–34

[22] Prakash S and Kumar S 2017 Experimental and theoretical analysis of defocused CO_2 laser microchanneling on PMMA for enhanced surface finish *J. Micromech. Microeng.* **27** 025003

[23] Prakash S and Kumar S 2017 Fabrication of rectangular cross-sectional microchannels on PMMA with a CO_2 laser and underwater fabricated copper mask *Opt. Laser Technol.* **94** 180–92

[24] De Marco C, Eaton S M, Suriano R, Turri S, Levi M, Ramponi R, Cerullo G and Osellame R 2010 Surface properties of femtosecond laser ablated PMMA *ACS Appl. Mater. Interfaces* **2** 2377–84

[25] Zielke D, Sylla D, Neubert T, Brendel R and Schmidt J 2013 Direct laser texturing for high-efficiency silicon solar cells *IEEE J. Photovolt.* **3** 656–61

[26] Archer M J and Ligler F S 2008 Fabrication and characterization of silicon micro-funnels and tapered micro-channels for stochastic sensing applications *Sensors* **8** 3848–72

[27] Walker C, Hungerford A, Narayanan G, Groppi C, Bloomstein T, Palmacci S, Stern M and Curtin J 1998 Laser micromachining of silicon: a new technique for fabricating terahertz imaging arrays *Proc. SPIE* **3357** 3357–8

[28] Conde J C, González P, Lusquiños F and León B 2009 Analysis of the formation and evolution of oriented microstructures on laser ablated silicon *Appl. Phys.* A **95** 465–71

[29] Chung C K, Wu M Y, Hsiao E J and Sung Y C 2007 Etching behavior of silicon using CO_2 laser *Proc. 2nd IEEE Int. Conf. on Nano/Micro Engineered and Molecular Systems* pp 59–62

[30] Lim D, Kamotani Y, Cho B, Mazumder J and Takayama S 2003 Fabrication of microfluidic mixers and artificial vasculatures using a high-brightness diode-pumped Nd:YAG laser direct write method *Lab Chip* **3** 318–23

[31] O'Neill W and Li K 2009 High-quality micromachining of silicon at 1064 nm using a high-brightness MOPA-based 20-w Yb fiber laser *IEEE J. Sel. Top. Quantum Electron.* **15** 462–70

[32] Chen T-C and Darling R B 2005 Parametric studies on pulsed near ultraviolet frequency tripled Nd:YAG laser micromachining of sapphire and silicon *J. Mater. Process. Technol.* **169** 214–8

[33] Kam D, Shah L and Mazumder J 2008 Laser based micro and nanopackaging and assembly *Proc. of SPIE* **6880** 68800s–8

[34] Kam D H, Shah L and Mazumder J 2011 Femtosecond laser machining of multi-depth microchannel networks onto silicon *J. Micromech. Microeng.* **21** 045027

[35] Ngoi B K A, Venkatakrishnan K, Lim L E N and Tan B 2001 Submicron micromachining on silicon wafer using femtosecond pulse laser *J. Laser Appl.* **13** 41–3

[36] Tran D, Zheng H, Lam Y, Murukeshan V, Chai J and Hardt D 2005 Femtosecond laser-induced damage morphologies of crystalline silicon by sub-threshold pulses *Opt. Lasers Eng.* **43** 977–86

[37] Amer M S, El-Ashry M, Dosser L, Hix K E, Maguire J F and Irwin B 2005 Femtosecond versus nanosecond laser machining: comparison of induced stresses and structural changes in silicon wafers *Appl. Surf. Sci.* **242** 162–7

[38] Wee L, Ng E, Prathama A and Zheng H 2011 Micro-machining of silicon wafer in air and under water *Opt. Laser Technol.* **43** 62–71

[39] Tangwarodomnukun V, Wang J, HuaHuang Z and Zhu H 2014 Heating and material removal process in hybrid laser-waterjet ablation of silicon substrates *Int. J. Mach. Tools Manuf.* **79** 1–16

[40] Li C-F, Johnson D and Kovacevic R 2003 Modeling of waterjet guided laser grooving of silicon *Int. J. Mach. Tools Manuf.* **43** 925–36

IOP Publishing

Laser Micro- and Nano-Scale Processing
Fundamentals and applications
Ahmed Issa and Dermot Brabazon

Chapter 11

Laser micro- and nano-processing: applications in modern dentistry

Yalda Afkham, Jennifer Gaughran, Verica Pavlic and Dermot Brabazon

Lasers have become irreplaceable devices in modern medicine and dentistry, covering a wide range of applications every day. Laser micro- and nano-processing are an important part of production and development (fabrication, integration, and assembly) of dental devices. They are also utilised as tools for diagnostics and treatments. Their usage has increased significantly given that laser devices are now much more inexpensive yet precise processing tools that can structure materials with a high degree of control, precision, and reduced residual impairment. However, there are still roadblocks for laser-based systems (such as temperature build-up effects, melting, burr formation, and cracking). Typically, only a limited number of materials are compatible with each laser device. The precise mechanism of laser irradiation and laser–biological material interaction is also not fully understood, which presents some barriers in their application.

11.1 Laser surface structuring

In the field of dentistry the interaction between cells and implant surfaces plays an essential role. The bone marrow mesenchymal stem cells adhere to the implant and become mature osteoblasts. This process is influenced by different events, such as the microtopography and the chemical composition of implants' surfaces. The ideal surface should be obtained through processes that do not modify its chemical composition maintaining the right degree of roughness and a cell-attractive microtopography [1–3].

The main advantage of using lasers within dentistry is their ability to fabricate a wide range of surface structures on a variety of metals and alloys, at the micro- and submicro-scale levels. Clearly, laser structuring of materials depends on a large number of parameters, related to the laser itself (the average laser power, pulse energy, pulse duration, repetition rate, wavelength etc), but also connected to specific sample properties such as type of material or its roughness [4]. Examination

doi:10.1088/978-0-7503-1683-5ch11

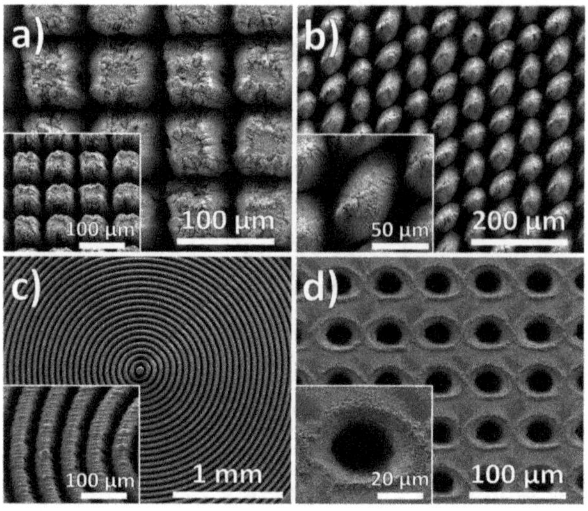

Figure 11.1. SEM micrograph patterns produced by laser processing Cu surfaces. (a) Square pillar, inset: sample tilted at 20° angle; (b) parallelogram structures in a hexagonal arrangement; (c) circular grooves; and (d) micro hole pattern. Reprinted from [4] with permission from MDPI.

of the influence of material topography on cellular behavior has shown that various structures such as pores, grooves and pits at the micrometer and nanometer dimensions influence cell morphology, adhesion, and proliferation of cells, leading to the definition of biologically favorable surfaces (figure 11.1).

In modern dentistry, namely, in implantology, the major issue relevant to the stability of implants is the fact that implants can withstand separation from the host tissue due to inadequate biocompatibility and poor implant osseointegration [5–10]. However, changing the surface chemistry and physical topography of the implant surface has been proven to influence biocompatibility and osseointegration. To date, several research studies report that the creation of nano- and microstructures on titanium implant material using lasers boosts the adherence of osteoblasts and helps fibroblasts to build up, which contributes to advanced implant osseointegration [5–10]. Laser processing is also effective in increasing the thickness of the surface oxide on titanium with a strong positive impact on bone–implant integration [9, 10]. Laser surface texturing can be seen therefore as is a relatively straightforward, flexible, precise, and affordable solution.

While titanium is a popular material for biomedical implants, several other materials, such as stainless steel, titanium/niobium/zirconium alloys, nickel/titanium alloy, and platinum [9–14], have been investigated and shown to result in similarly beneficial properties for biocompatibility. Apart from osteoblasts and fibroblasts, selective proliferation of neurons was demonstrated on a silicon surface micro-structured by a femtosecond laser [11]. The improved osseointegration is correlated with superhydrophobic surface structures, also known as the lotus leaf effect [15]. In nature, the lotus leaf has a superhydrophobic surface with self-cleaning effects [15–19]. Large varieties of nano- and microstructures that result in superhydrophobic surfaces have been created on different metal surfaces by femtosecond laser micromachining [15–19]. Moreover, Kietzig *et al* noted that it takes some time for superhydrophobicity to develop on intrinsically hydrophilic metals, which they attributed to the formation of a carbonaceous surface layer after laser treatment that alters the chemistry of the surface [18].

Further, in modern dentistry, lasers on micro- and submicrometer levels have been proposed as an alternative effective way for treating zirconia surfaces in an attempt to improve the adhesion of dental cements and orthodontic brackets. This is achieved by etching the surface with precision, without producing mechanical degradation of the materials and without raising the temperature of the irradiated surface [20].

Laser machining of biocompatible materials has also been used to create scaffolds for tissue engineering with controlled pore size and porosity, and furthermore with controlled cell orientation and location. In the most commonly used polymers for tissue engineering, laser micro- and nanomachining can be used to improve the proliferation of cells within interior regions of these materials [21]. The implementation of trial laser parameters for laser micromachining can be taken from those already implemented from the production of microchannels in microfluidic devices [22, 23].

11.2 Laser tissue bonding

To date, the most desirable wound healing is healing by primary intention (i.e. the healing of a clean wound without tissue loss). In clinical practice, it is achieved via surgical sutures, staples and clips. However, these methods can cause a foreign body reaction due to the nature of the materials used, resulting in inflammation, granuloma formation, and scarring [24, 25]. Regardless of the methods used, an impermeable seal over the repair is rarely obtained. Furthermore, traditional methods are almost inapplicable to tissue closure in microsurgery such as vascular anastomosis and closure of nerves [26]. For these reasons new methods of wound closure have been sought and introduced, such as laser tissue bonding (LTB). LTB is a minimally invasive technique that can be faster, nonimmunogenic (does not produce an immune response), less traumatic, and easier to apply than conventional tissue closure methods under optimum circumstances [25]. To date, LTB has been used to repair a variety of tissues, such as skin, liver, cartilage, urinary tract and nerves, as well as vascular anastomoses [27]. LTB is divided into three types, namely, laser tissue welding, laser tissue soldering (LTS), and dye-enhanced LTS, which are shown in figure 11.2.

Laser tissue welding is a surgical technique which uses a laser energy to join or bond tissues (photothermal and/or photochemical bonds). In general, this energy results in some alteration of the molecular structure of the tissues being joined [26].

Laser tissue soldering is an alternative approach for tissue bonding. This technique is based on photo-enhancement by applying some soldering material (liquid and semi-solid biological glues based on proteins and other compounds) onto the approximated edges of the cut. The solder and the underlying tissues are then

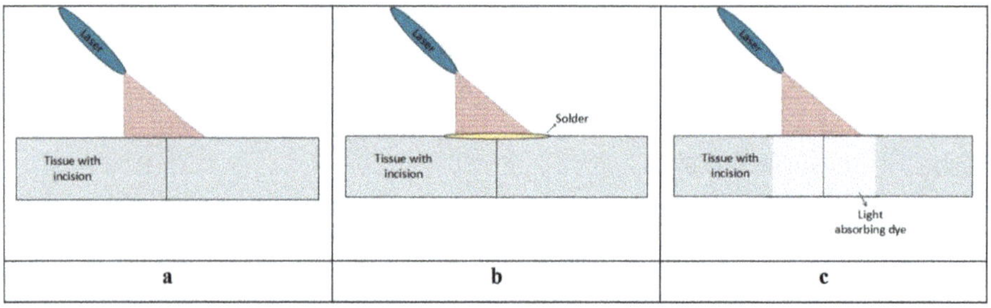

Figure 11.2. Schematics of tissue repair using: (a) laser welding, (b) laser soldering, and (c) dye-enhanced laser soldering.

heated by a laser light. Since laser tissue soldering does not involve a foreign body, it offers some advantages over conventional techniques such as watertight seals, and faster and scarless healing [28]. Furthermore, laser solders can include chromophores that are used to control the laser penetration. A known concentration of absorbing chromophore is added to the solder in order to focus light absorption in the solder itself, without changing the native tissue or body fluids. This technique allows the use of a lower power density, known to be safer for use (less of an eye safety hazard), smaller in size and less expensive, thus, resulting in reduced collateral thermal damage [29, 30]. For optimum results, the combination of tightly controlled laser exposure parameters and thermal diffusion characteristics of the solder gives predictable and reproducible tissue welding [31]. In the early stages of this process, the introduction of fibrinogen-based solders, like cryoprecipitate in humans, were used but this raised concerns over infection risks, stability, and handling properties [29]. Since then, typical additives trialed have included native collagen, gelatinous collagen, fibrin, elastin, and albumin [26].

Regardless of the choice of solder, generally more energy is absorbed near the upper portion of the solder, near the laser spot. A temperature gradient is established over the depth of the solder. Depending on the temperature gradient and the laser exposure, the upper portion of the solder can become overcoagulated, while the most critical region, the solder/tissue interface, does not get fully coagulated. Such undercoagulated solder has been shown to create unstable bonds. Temperature-controlled laser soldering offers an accelerated wound reparative process with numerous advantages over the conventional methods [32].

Dye-enhanced laser tissue soldering is a more recent soldering technique, where laser tissue welding is improved by use of solder and photodynamic therapy (laser energy and wavelength specific absorbing dye). Most recently, it has been successfully applied both experimentally and clinically in the bowel, urethra, ureter, and blood vessels with excellent results. With this method a rapid and safe watertight seal can be achieved, with minimal foreign body reaction [33].

A main issue that needs attention within the LTB processes is the potential for thermal damage to the surrounding tissue. The available literature suggests that lasers with powers of 0.5–1 W, even with typical small spot sizes, result in imperceptible thermal damage [32]. The maximum tissue temperature is effected by different laser process parameters, wavelengths, and spot sizes being used. To overcome these disadvantages, laser wavelengths within the near infrared spectrum are used [27]. This means that wavelengths which are highly absorbed by water and with shallow tissue penetration depths are recommended to be used. In this case, the depth of thermal damage can be restricted to the most superficial adventitia (the outer layer of the blood vessel wall).

Precise temperature control is a crucial issue in laser bonding and has led to the development of more refined welding techniques, such as temperature-controlled tissue laser soldering and/or the use of temperature feedback control systems. This innovative approach is awaiting solid experimental data to become the gold-standard surgical procedure for incision closure [34].

Although many researchers agree that laser tissue welding is a promising new tissue welding technique in clinical practice, major gaps in basic understanding of the actual mechanism involved in the process prevent it from being introduced into everyday practice. Furthermore, the fact that different researchers are using significantly different laser irradiation parameters, different laser wavelengths, and different tissues to be welded in their research, makes it difficult to generate a single laser system that can be used broadly for different tissues.

11.3 Additive manufacturing (3D printing) of dental implants

The term additive manufacturing (3D printing) is generally used to describe a manufacturing approach in which objects are built one layer at a time, adding multiple layers to form an object. The low cost of manufacturing and less material waste are noted advantages of using additive manufacturing technology [35]. It has revolutionized medicine and dentistry. In comparison to restoration done by dental technicians, restorative dentistry using additive manufacturing can provide shorter processing time, reduced overall costs, improved availability, and allows for the printing of items with complex structures [36]. Additive manufacturing classification includes stereolithography (SLA), selective laser sintering (SLS), fused deposition modeling (FDM), selective laser melting (SLM), ink-jetting, and electronic beam melting (EBM). Also, it can be classified as liquid-based, solid-based, or powder-based materials. The advantages and disadvantages of different additive manufacturing techniques are listed in table 11.1. Table 11.2 also indicates the material used in each technique and their potential applications in dentistry [33, 37–41].

SLA is an additive manufacturing process which originated as a vat photo-polymerization technique. In this method an object is created by selectively curing a liquid polymer resin layer-by-layer using an ultraviolet (UV) laser beam. In spite of the fact that this method is a high cost process for making large objects, it is widely used in the production of 3D printed implant drill guides [42, 43].

In-jetting printers usually consist of two or more jetting heads. One head is used to jet out build material, and the other head is used to jet out support material. Using multiple print heads is a comparative advantage of this technology that allows concurrent printing with different materials. Adjusting the mixture of materials allows for the production of various properties within the printed object which can

Table 11.1. Additive manufacturing techniques.

Techniques	Advantages	Disadvantages
Stereolithography	• Rapid fabrication • Customized coloring • Able to create complex shapes with high feature resolution • Low material consumption	• Depending on the material, components may be brittle • High cost technology • Support materials must be removed • Photosensitive resin is a problem • Limited shelf life and vat life. Cannot be heat sterilized
Jetting	• Relatively fast • Lower cost technology • Excellent resolution • High quality • Rigid, opaque, multi-color, transparent and flexible materials	• Tenacious support material can be difficult to remove completely • Support material may cause skin irritation • Cannot be heat sterilized • High cost materials
Selective laser sintering	• Range of polymeric materials including nylon, elastomers, and composites • Fast • Excellent layer adhesion • Strong and accurate parts • No need for support structures • Polymeric materials—commonly nylon may be autoclaved • Printed object may have full mechanical functionality • Lower cost materials	• Significant infrastructure required • Produces a lot of waste • Lower cost in bulk • Inhalation risk • Messy • Expensive • Produces rough surfaces
Selective laser melting	• High mechanical load capacity • Variety of materials including titanium, titanium alloys, cobalt chrome, and stainless steel • Almost no restriction on the shape of the product • Shorter assembly times • Metal alloy may be recycled • Does not require expensive production equipment	• Elaborate infrastructure requirements • High initial costs • Dust and nanoparticle condensate may be hazardous to health • Acute size restrictions • Explosive risk • Produces rough surfaces • Elaborate post-processing is required • Hard to remove support materials • Relatively slow process
Electronic beam melting	• Residual stress reduction due to increased process temperature • Minimum material waste • High speed	• Limited commercially available materials • Dust may be hazardous to health • Explosive risk • Produces rough surfaces

	• Flexible process without tooling and set-up costs • Dense parts with controlled porosity • Geometric freedom for engineering product designers	• Limited build size • Less post-processing required • Lower resolution
Fused deposition modeling	• Less time consuming • High porosity • Simple to use • Variable mechanical strength • Low to mid-range cost materials and equipment • Low accuracy in low cost equipment • Broad range in materials	• Low cost but limited materials—only thermoplastics • Limited shape complexity for biological materials • Support material must be removed

Table 11.2. 3D printers, the materials used and their application in dentistry.

Techniques	Materials	Application in dentistry
Stereolithography	A wide variety of resins such as high temperature resin, dental resin, castable resin, etc	Dental replica models, surgical guides and splints, orthodontic devices, temporary and definitive crowns, temporary bridges
Jetting	A wide variety of photopolymers	Orthodontic models, implantology case planning, cast partial frames and sophisticated anatomical models
Selective laser sintering and selective laser melting	Powder such as alumide, polyamide, glass-particle filled polyamide, rubber-like polyurethane, etc; stainless steel, cobalt chrome, titanium alloy	Hospital set-up for metal crowns, copings and bridges, metal or resin partial denture frameworks
Fused deposition modeling	Thermoplastic polymers such as polylactic acid, acrylonitrile butadiene styrene, nylon, polyethylene terephthalate glycol-modified, polycarbonate, polyether ether ketone, etc	Custom impressions, custom bite registrations, wax set-up for tray-in

affect, for example, the flexibility of the object. It can, for instance, be used in production of indirect orthodontic bracket splints. A linear actuator moves the print heads in the X–Y plane above a build plate and a UV lamp located next to the print heads instantly cures the droplets. For the next layer to be printed, the build plate moves down by an increment. The thickness of each layer is usually between 14 and 28 μm to achieve a fine level of detail [44–46].

Selective laser sintering (SLS) technology has been available since the 1980s. This technique employs a high-energy laser beam to heat and fuse polymer particles into complex, net-shaped, 3D components as the laser beam repeatedly scans over a single layer of powder granules and consolidates them via full or partial melting. This method results in a high level of precision and since the printed structure is supported by surrounding powder, further support material is not required. [47–49]. As a rapid and accurate process, it allows straightforward batch production of implants for orthopedic and dental applications. It is also considered a leading technology in the production of titanium cranioplasties for oral and maxillofacial surgery [50–53].

Selective laser melting (SLM) is a specific 3D printing technique, which utilizes a high power-density laser to fully melt and fuse metallic powders and produce near net-shape parts with near full density. SLM is one of the most exciting technologies available today and is utilized both for rapid prototyping and mass production. The range of metal alloys available is fairly extensive. Via process parameter control, the properties of parts produced in this manner can be made to a similar standard to those manufactured via traditional manufacturing processes. SLM is similar to SLS in that both processes are covered under the powder bed fusion umbrella. The major difference is the type of feedstock or powder being used; while SLS uses polymer materials, metal powders are used with SLM. Aside from this, due to constraints of the SLM process and the weight of the material, SLM may require support structures for overhanging features. This is a significant difference from SLS, where in SLS the surrounding powder material can provide enough support, allowing freeform shapes and features to be readily realized [48, 54].

Electron beam melting (EBM) is used to manufacture parts by melting the metal powder layer-by-layer using a high power electron beam in a high vacuum. The electron beam is used to provide the energy needed for a high level of melting capacity and productivity. The process is often performed at an evaluated temperature thus allowing parts to be produced with no or low levels of residual stress. The vacuum also ensures a clean and controlled environment. EBM and SLM technology offers good freedom of design with higher volume possible due to its ability to tightly stack parts. A combination that allows for the manufacture of complex and detailed orthopedic implants [55, 56].

Fused deposition modeling (FDM) printers are more common in the medical and dental industry due to their widespread availability, ease of installation, and economic affordability. In this technique, the material is supplied into a hot nozzle which melts and extrudes material in two dimensions (X–Y). Notwithstanding the fact that it permits printing of relatively low resolution anatomical models with less complexity, it is less capable for printing complex anatomies due to limitations of color selection, resolution, and the requirement for removal of support material [57, 58].

In order to fabricate 3D printed biomodels, digital imaging such as intraoral scanning, cone beam computed tomography (CBCT), and magnetic resonance imaging (MRI) have been used together with 3D printing [59]. CBCT provides volumetric data for 3D printing [33]. It has become more popular due to the advantages of CBCT over conventional radiography and other three-dimensional imaging modalities such as its high-resolution imaging procedure in oral and maxillofacial radiology [60]. CBCT has a 2D sensor and instead of a fan-shaped beam (as is used in CT) it uses a cone-shaped X-ray beam. By a single 360° rotation of the beam and sensor around the skull, volume data can be acquired [61, 62]. In comparison with conventional CT, CBCT provides higher resolution at lower radiation doses [63, 64]. This revolutionary technology is widely used in different areas of dentistry, especially in orthodontics, restorative dentistry, prosthodontics and maxillofacial, and oral and implant surgery [65].

Nowadays, with the advent of 3D printing technology, scanning and modeling of dental casts are possible. In order to correct malpositioned teeth and jaws, orthodontists use 3D software to customize the treatment plan via a virtual 3D interface. This can include algorithms to calculate the amount of force needed on an individual tooth to achieve the desired movement [66, 67].

In surgery, a large amount of lost tissues (tumor- or trauma-related tissue loss) can be replaced by detailed 3D replicas [68]. It is a less invasive approach that leads to more efficient, affordable, and predictable surgery for patients. Above all, it allows dentists to create patient-tailored dentistry or personalized dentistry, which is an imperative of modern dentistry, especially in dental surgery [33]. Interestingly, the use of laser devices delivering nanosecond pulses have also been shown to be able to induce sterilization or decontamination of the biomaterials, extending the usage of biomaterials and surgical instruments [11, 69].

As a wide range of dental materials such as thermoplastic polymers, waxes, metals and alloys, ceramics and thermoplastic composites can be utilized for manufacturing dental constructions, as well as the need for complex geometries and customization capability of the dental prostheses, SLS/SLM technology becomes suitable for application in dental medicine. Using SLS, the maxilla-facial prostheses, functional skeletons and individual scaffolds for tissue engineering can be fabricated of polymers and composites. When the metals and alloys are processed by SLM, bulk as well as porous orthopedic and dental implants, dental crowns, bridges and frameworks for partial prostheses can be produced [70–72].

SLS/SLM technology is suitable for application in dental medicine as it can be utilized on an entire range of dental materials such as thermoplastic polymers, waxes, metals and alloys, ceramics and thermoplastic composites for manufacturing dental constructions, also for its ability to manufacture complex geometries and for its customization capability of the dental prostheses. Using SLS, the maxilla-facial prostheses, functional skeletons and individual scaffolds for tissue engineering can be fabricated of polymers and composites. When the metals and alloys are processed by SLM, bulk as well as porous orthopedic and dental implants, dental crowns, bridges and frameworks for partial prostheses can be produced [73–75].

In medical applications, high porosity of implants is one of the most important factors since it improves the attachments between muscle and implant. A fully digital method for designing and manufacturing of personal frameworks for complex dental prostheses has been developed by SLM of titanium or cobalt–chromium. The framework is the metal base structure of the prosthesis and supports the artificial teeth (figure 11.3) [76–78].

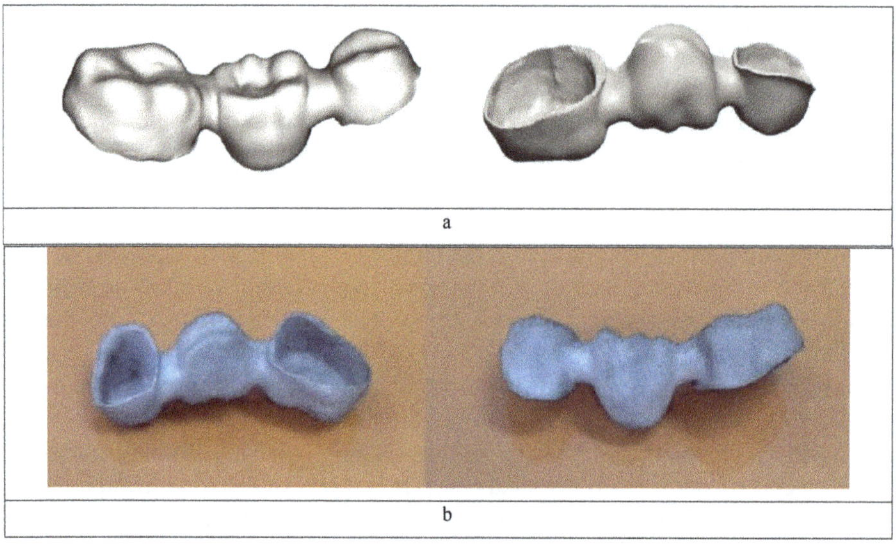

Figure 11.3. (a) 3D CAD models of a dental bridge, and (b) SLM-manufactured dental bridge. Reprinted from [78] with permission from MDPI.

In comparison with cast alloys, selective laser melting of cobalt–chromium alloys have high mechanical and tribocorrosion properties, comparatively good fitting ability, and higher adhesion strength of the porcelain. All this is a good precondition for successful application of the SLM process in the production of fixed dental prostheses, mainly of frameworks for metal–ceramic and constructions covered with polymer-composite, intended for areas with high loading [79, 80].

11.4 Conclusion

Additive manufacturing has attracted the interest of researchers in various applications and fields. In dentistry, computerization has come to the fore with the introduction of 3D imaging, modeling and CAD technologies, which are hugely impacting all aspects in this market. 3D printing makes it possible to accurately make one-off, complex geometrical forms from digital data, in a variety of materials, in local or large industrial centers. Nowadays, the main focus is on surgical planning and the indirect production of implants or orthodontic aligners by printing the molds for them. A suitable printing system is normally selected by considering availability and medical properties of the material, the time required and the desired resolution [81, 82]. Based on the findings of reviews, the following conclusions can be noted:

- Additive manufacturing is an effective alternative for the manufacture of custom implants.
- 3D printing techniques and materials should be chosen based on the required accuracy, strength and biocompatibility of the orthodontic appliance.
- This technique is a cost-effective fabrication method for dental implants and can allow better osseointegration.

- Additive manufacturing of implants brings advantages such as customization, flexibility, and freedom in design and enables also manipulation of chemical and physical properties.
- Obtaining the required surface quality and dimensional accuracy still remain a main challenge for the advancement of some additive manufacturing technologies for dental implants.

Acknowledgments

This work is supported by a research grant from Science Foundation Ireland (SFI) under Grant Number 16/RC/3872 and is co-funded under the European Regional Development Fund and by I-Form industry partners.

References

[1] Conserva E, Consolo U and Bellini P 2018 Adhesion and proliferation of human dental pulp stem cells on a laser microtextured implant surface: an *in vitro* study *Oral Health and Care* **25** 26

[2] Feller L, Jadwat Y, Khammissa R A, Meyerov R, Schechter I and Lemmer J 2015 Cellular responses evoked by different surface characteristics of intraosseous titanium implants *BioMed Res. Int.* **2015** 1–8

[3] Conserva E, Menini M, Ravera G and Pera P 2013 The role of surface implant treatments on the biological behavior of SaOS-2 osteoblast-like cells. An *in vitro* comparative study *Clin. Oral Implants Res.* **24** 880–9

[4] Ahmmed K, Grambow C and Kietzig A-M 2014 Fabrication of micro/nano structures on metals by femtosecond laser micromachining *Micromachines* **5** 1219–53

[5] Hirao M *et al* 2005 Macro-structural effect of metal surfaces treated using computer-assisted yttrium–aluminum–garnet laser scanning on bone-implant fixation *J. Biomed. Mater. Res.* A **73** 213–22

[6] Müller M, Hennig F, Hothorn T and Stangl R 2006 Bone–implant interface shear modulus and ultimate stress in a transcortical rabbit model of open-pore Ti6Al4V implants *J. Biomech.* **39** 2123–32

[7] Gaggl A, Schultes G, Müller W and Kärcher H 2000 Scanning electron microscopical analysis of laser-treated titanium implant surfaces—a comparative study *Biomaterials* **21** 1067–73

[8] Hallgren C, Reimers H, Chakarov D, Gold J and Wennerberg A 2003 An *in vivo* study of bone response to implants topographically modified by laser micromachining *Biomaterials* **24** 701–10

[9] Palmquist A, Lindberg F, Emanuelsson L, Brånemark R, Engqvist H and Thomsen P 2010 Biomechanical, histological, and ultrastructural analyses of laser micro- and nano-structured titanium alloy implants: a study in rabbit *J. Biomed. Mater. Res.* A **92** 1476–86

[10] Forsgren J, Paz M D, León B and Engqvist H 2013 Laser induced surface structuring and ion conversion in the surface oxide of titanium: possible implications for the wetability of laser treated implants *J. Mater. Sci., Mater. Med.* **24** 11–5

[11] Kenar H *et al* 2013 Femtosecond laser treatment of 316L improves its surface nanoroughness and carbon content and promotes osseointegration: an *in vitro* evaluation *Colloids Surf.* B **108** 305–12

[12] Jeong Y-H, Choe H-C and Brantley W A 2011 Nanostructured thin film formation on femtosecond laser-textured Ti–35Nb–*x*Zr alloy for biomedical applications *Thin Solid Films* **519** 4668–75
[13] Ranella A, Barberoglou M, Bakogianni S, Fotakis C and Stratakis E 2010 Tuning cell adhesion by controlling the roughness and wettability of 3D micro/nano silicon structures *Acta Biomater.* **6** 2711–20
[14] Aguilar C A, Lu Y, Mao S and Chen S 2005 Direct micro-patterning of biodegradable polymers using ultraviolet and femtosecond lasers *Biomaterials* **26** 7642–9
[15] Zorba V *et al* 2008 Biomimetic artificial surfaces quantitatively reproduce the water repellency of a lotus leaf *Adv. Mater.* **20** 4049–54
[16] Moradi S, Kamal S, Englezos P and Hatzikiriakos S G 2013 Femtosecond laser irradiation of metallic surfaces: effects of laser parameters on superhydrophobicity *Nanotechnology* **24** 415302
[17] Wu B, Zhou M, Li J, Ye X, Li G and Cai L 2009 Superhydrophobic surfaces fabricated by microstructuring of stainless steel using a femtosecond laser *Appl. Surf. Sci.* **256** 61–6
[18] Kietzig A-M, Hatzikiriakos S G and Englezos P 2009 Patterned superhydrophobic metallic surfaces *Langmuir* **25** 4821–7
[19] Baldacchini T, Carey J E, Zhou M and Mazur E 2006 Superhydrophobic surfaces prepared by microstructuring of silicon using a femtosecond laser *Langmuir* **22** 4917–9
[20] García-Sanz V *et al* 2017 Effects of femtosecond laser and other surface treatments on the bond strength of metallic and ceramic orthodontic brackets to zirconia *PLoS One* **12** 1–11
[21] Lannutti J, Reneker D, Ma T, Tomasko D and Farson D 2007 Electrospinning for tissue engineering scaffolds *Mater. Sci. Eng.* C **27** 504–9
[22] Ke K, Hasselbrink E F and Hunt A J 2005 Rapidly prototyped three-dimensional nano-fluidic channel networks in glass substrates *Anal. Chem.* **77** 5083–8
[23] Kim T N, Campbell K, Groisman A, Kleinfeld D and Schaffer C B 2005 Femtosecond laser-drilled capillary integrated into a microfluidic device *Appl. Phys. Lett.* **86** 201106
[24] Mistry Y A, Natarajan S S and Ahuja S A 2018 Evaluation of laser tissue welding and laser-tissue soldering for mucosal and vascular repair *Ann. Maxillofac. Surg.* **8** 35
[25] Bass L S and Treat M R 1995 Laser tissue welding: a comprehensive review of current and future *Lasers Surg. Med.* **17** 315–49
[26] Kim K and Guo Z 2004 Ultrafast radiation heat transfer in laser tissue welding and soldering *Numer. Heat Transfer* A **46** 23–40
[27] Gobin A M, O'Neal D P, Watkins D M, Halas N J, Drezek R A and West J L 2005 Near infrared laser-tissue welding using nanoshells as an exogenous absorber *Lasers Surg. Med.* **37** 123–9
[28] Forer B *et al* 2007 CO_2 laser fascia to dura soldering for pig dural defect reconstruction *Skull Base* **17** 17–23
[29] Oz M C *et al* 1990 Indocyanine green dye enhanced vascular welding with the near infrared diode laser *Vasc. Surg.* **24** 564–70
[30] Oz M C *et al* 1990 Tissue soldering by use of indocyanine green dye-enhanced fibrinogen with the near infrared diode laser *J. Vasc. Surg.* **11** 718–25
[31] Simhon D *et al* 2004 Closure of skin incisions in rabbits by laser soldering: I: Wound healing pattern *Lasers Surg. Med.* **35** 1–11
[32] Perveen A, Molardi C and Fornaini C 2018 Applications of laser welding in dentistry: a state-of-the-art review *Micromachines* **9** 209

[33] Dawood A, Marti B M, Sauret-Jackson V and Darwood A 2015 3D printing in dentistry *Br. Dental J.* **219** 521
[34] Small W IV 1998 Thermal and molecular investigation of laser tissue welding *PhD Thesis* Lawrence Livermore National Lab
[35] Gibson I, Rosen D W and Stucker B 2014 *Additive Manufacturing Technologies* (Berlin: Springer)
[36] Zaharia C *et al* 2017 Digital dentistry—3D printing applications *J. Interdiscipl. Med.* **2** 50–3
[37] Jardini A L *et al* 2014 Cranial reconstruction: 3D biomodel and custom-built implant created using additive manufacturing *J. Cranio Maxillof. Surg.* **42** 1877–84
[38] Milovanović J and Trajanović M 2007 Medical applications of rapid prototyping *Facta Univ. Mech. Eng.* **5** 79–85
[39] Chae M P, Rozen W M, McMenamin P G, Findlay M W, Spychal R T and Hunter-Smith D J 2015 Emerging applications of bedside 3D printing in plastic surgery *Front. Surg.* **2** 25
[40] Rengier F *et al* 2010 3D printing based on imaging data: review of medical applications *Int. J. Comput. Assisted Radiol. Surg.* **5** 335–41
[41] Oberoi G, Nitsch S, Edelmayer M, Janjić K, Müller A S and Agis H 2018 3D printing—encompassing the facets of dentistry *Front. Bioeng. Biotechnol.* **6** 172
[42] Prakash K S, Nancharaih T and Rao V S 2018 Additive manufacturing techniques in manufacturing–an overview *Mater. Today Proc.* **5** 3873–82
[43] Dehurtevent M, Robberecht L, Hornez J-C, Thuault A, Deveaux E and Béhin P 2017 Stereolithography: a new method for processing dental ceramics by additive computer-aided manufacturing *Dent. Mater.* **33** 477–85
[44] Ibrahim D *et al* 2009 Dimensional error of selective laser sintering, three-dimensional printing and PolyJet™ models in the reproduction of mandibular anatomy *J. Cranio Maxillof. Surg.* **37** 167–73
[45] Anderson J, Wealleans J and Ray J 2018 Endodontic applications of 3D printing *Int. Endodontic J.* **51** 1005–18
[46] Brown G B, Currier G F, Kadioglu O and Kierl J P 2018 Accuracy of 3-dimensional printed dental models reconstructed from digital intraoral impressions *Am. J. Orthod. Dentofacial Orthop.* **154** 733–9
[47] Deckard C R 1989 Method and apparatus for producing parts by selective sintering *US Patent* US4863538A
[48] Kruth J-P, Vandenbroucke B, Van Vaerenbergh J and Mercelis P 2005 Benchmarking of different SLS/SLM processes as rapid manufacturing techniques *PMI 2005: Proc. of the Int. Conf. Polymers & Moulds Innovations (PMI) (Gent, Belgium, April 20–23)*
[49] Kalsoom U *et al* 2018 Low-cost passive sampling device with integrated porous membrane produced using multimaterial 3D printing *Anal. Chem.* **90** 12081–9
[50] Takemoto M *et al* 2016 Additive-manufactured patient-specific titanium templates for thoracic pedicle screw placement: novel design with reduced contact area *Eur. Spine. J.* **25** 1698–705
[51] Schepers R H *et al* 2013 Fully 3-dimensional digitally planned reconstruction of a mandible with a free vascularized fibula and immediate placement of an implant-supported prosthetic construction *Head Neck* **35** E109–14
[52] Lin W-S, Starr T L, Harris B T, Zandinejad A and Morton D 2013 Additive manufacturing technology (direct metal laser sintering) as a novel approach to fabricate functionally graded titanium implants: preliminary investigation of fabrication parameters *Int. J. Oral Maxillofac. Implants* **28** 1490–5

[53] Tsai M-J and Wu C-T 2014 Study of mandible reconstruction using a fibula flap with application of additive manufacturing technology *Biomed. Eng. Online* **13** 57

[54] Vandenbroucke B and Kruth J P 2007 Selective laser melting of biocompatible metals for rapid manufacturing of medical parts *Rapid Prototyp. J.* **13** 196–203

[55] Koike M, Martinez K, Guo L, Chahine G, Kovacevic R and Okabe T 2011 Evaluation of titanium alloy fabricated using electron beam melting system for dental applications *J. Mater. Process. Technol.* **211** 1400–8

[56] Ataee A, Li Y, Fraser D, Song G and Wen C 2018 Anisotropic Ti–6Al–4V gyroid scaffolds manufactured by electron beam melting (EBM) for bone implant applications *Mater. Des.* **137** 345–54

[57] Huang Y, Zhang X F, Gao G, Yonezawa T and Cui X 2017 3D bioprinting and the current applications in tissue engineering *Biotechnol. J.* **12** 1600734

[58] Melchels F P, Feijen J and Grijpma D W 2010 A review on stereolithography and its applications in biomedical engineering *Biomaterials* **31** 6121–30

[59] Suomalainen A, Stoor P, Mesimäki K and Kontio R K 2015 Rapid prototyping modelling in oral and maxillofacial surgery: a two year retrospective study *J. Clin. Exp. Dentist* **7** e605

[60] Hegde S, Ajila V, Kamath J S, Babu S, Pillai D S and Nair S M 2018 Importance of cone-beam computed tomography in dentistry: an update *SRM J. Res. Dental Sci.* **9** 186

[61] Toyofuku F, Konishi K and Kanda S 1986 Representation of arbitrarily curved sections of dentomaxillofacial region by the *X*-ray video CT *Oral Radiol.* **2** 9–13

[62] Kiljunen T, Kaasalainen T, Suomalainen A and Kortesniemi M 2015 Dental cone beam CT: A review *Phys. Med.* **31** 844–60

[63] Arai Y, Tammisalo E, Iwai K, Hashimoto K and Shinoda K 1999 Development of a compact computed tomographic apparatus for dental use *Dentomaxillofac. Radiol.* **28** 245–8

[64] Honda K 1998 Clinical experience with ortho-CT for the diagnosis of the temporomandibular joint disorders *Dentomaxillofac. Radiol.* **27** 39

[65] Kim S-Y, Shin Y-S, Jung H-D, Hwang C-J, Baik H-S and Cha J-Y 2018 Precision and trueness of dental models manufactured with different 3-dimensional printing techniques *Am. J. Orthod. Dentofac. Orthop.* **153** 144–53

[66] Vasamsetty P, Pss T, Kukkala D, Singamshetty M and Gajula S 2020 3D printing in dentistry—exploring the new horizons *Mater. Today Proc.* **26** 838–41

[67] Kasparova M *et al* 2013 Possibility of reconstruction of dental plaster cast from 3D digital study models *Biomed. Eng. Online* **12** 49

[68] Owusu J A and Boahene K 2015 Update of patient-specific maxillofacial implant *Curr. Opin. Otolaryngol. Head Neck Surg.* **23** 261–4

[69] Valette S, Steyer P, Richard L, Forest B, Donnet C and Audouard E 2006 Influence of femtosecond laser marking on the corrosion resistance of stainless steels *Appl. Surf. Sci.* **252** 4696–701

[70] Kruth J-P, Mercelis P, Vaerenbergh J V, Froyen L and Rombouts M 2005 Binding mechanisms in selective laser sintering and selective laser melting *Rapid Prototyp. J.* **11** 26–36

[71] Furumoto T *et al* 2015 Permeability and strength of a porous metal structure fabricated by additive manufacturing *J. Mater. Process. Technol.* **219** 10–6

[72] Tara M A, Eschbach S, Bohlsen F and Kern M 2011 Clinical outcome of metal-ceramic crowns fabricated with laser-sintering technology *Int. J. Prosthod.* **24** 46–8

[73] Azari A and Nikzad S 2009 The evolution of rapid prototyping in dentistry: a review *Rapid Prototyp. J.* **15** 216–25

[74] Andonović V and Vrtanoski G 2010 Growing rapid prototyping as a technology in dental medicine *Mech. Eng. Sci. J.* **29** 31–9

[75] McAlea K, Forderhase P, Hejmadi U and Nelson C 1997 Materials and applications for the selective laser sintering *Proc. 7th Int. Conf. on Rapid Prototyping (San Francisco, CA)* pp 23–33

[76] Dobrzański L A, Dobrzańska-Danikiewicz A D, Achtelik-Franczak A, Dobrzański L B, Szindler M and Gaweł T G 2017 Porous selective laser melted Ti and Ti6Al4V materials for medical applications *Powder Metallurgy–Fundamentals and Case Studies* (Rijeka: InTech) pp 161–81

[77] Cosma C 2012 Manufacturing of implants by selective laser melting *Balneo-Res. J.* **3** 1–18

[78] Cosma C, Kessler J, Gebhardt A, Campbell I and Balc N 2020 Improving the mechanical strength of dental applications and lattice structures SLM processed *Materials* **13** 905

[79] Dzhendov D and Dikova T 2016 Application of selective laser melting in manufacturing of fixed dental prostheses *J. IMAB* **22** 1414–7

[80] Torabi K, Farjood E and Hamedani S 2015 Rapid prototyping technologies and their applications in prosthodontics, a review of literature *J. Dent.* **16** 1

[81] Werz S M, Zeichner S, Berg B I, Zeilhofer H F and Thieringer F 2018 3D printed surgical simulation models as educational tool by maxillofacial surgeons *Eur. J. Dental Educ.* **22** e500–5

[82] Torres K, Staśkiewicz G, Śnieżyński M, Drop A and Maciejewski R 2011 Application of rapid prototyping techniques for modelling of anatomical structures in medical training and education *Folia morphologica* **70** 1–4

www.ingramcontent.com/pod-product-compliance
Ingram Content Group UK Ltd.
Pitfield, Milton Keynes, MK11 3LW, UK
UKHW051854100926
8556IPUK00005B/616

9 780750 316811